MACMILLAN · FIELD · GUIDES

·FOSSILS·

The author

Dr. Richard Moody is Senior Lecturer in Paleontology at Kingston Polytechnic, Surrey, England. His books include *A Natural History of Dinosaurs*, *Prehistoric Life*, and *Evolution of Life*.

Acknowledgements

Photographs

All the photographs in this book were supplied by RIDA Photo Library, with the exception of the following: page 12 (Novosti Press Agency).
Identification plates by RIDA Photo Library
Illustrations by Format Publishing Services
Edited and designed by Format Publishing Services.

Copyright © 1986, Hamlyn Publishing,
a division of The Hamlyn Publishing Group Limited.
This Edition published 1986 by Macmillan Publishing Company,
a division of Macmillan, Inc.

Macmillan Publishing Company
866 Third Avenue, New York, N.Y. 10022
Collier Macmillan Canada, Inc.

Library of Congress Cataloging-in-Publication Data

Moody, Richard.
 Fossils.

 (Macmillan field guides)
 1. Paleontology – Indentification. I. Title
II. Series
QE718.M65   1986   560   85–21372

 ISBN 0–02–063370–X

10 9 8 7 6 5 4 3 2 1

Printed in Italy

MACMILLAN · FIELD · GUIDES

· FOSSILS ·

RICHARD MOODY

Collier Books

Macmillan Publishing Company

New York

Foreword

This book is a practical fieldguide for both interested amateurs and students at school or college. With it, the reader should not only be able to understand something of the science of palaeontology (the study of the remains of plants and animals), but also how to go about finding and identifying fossils in the field. Fieldwork is an important element in the retrieval of data, and the discovery of new species still awaits the amateur even today. This book also explains the way in which fossils are formed and the techniques used in their collection and preservation. Specimens collected in the field help in our understanding of biological evolution and the changes which have taken place in animal and plant communities over eons of time, and each specimen found helps to piece together the jigsaw.

R.T.M.

Contents

Introduction

Geology is a science of many parts, and most geologists are specialists who tend to restrict their research interests. Those who study the nature and composition of rocks are called petrologists, while others who observe, measure and record folds and associated features are known as structural geologists. Stratigraphers unravel the history and relationships of rocks deposited throughout the course of time, whereas palaeontologists identify and describe the remains of animals and plants that lived thousands, millions or even billions of years ago. Such remains are termed fossils, and they are the subject of this book. This book will help you collect and identify your own fossils. It will also help you understand and interpret their life styles and encourage personal comments on the relationships between the fossils and the sediments in which they were buried. For the geologist, the term **field** refers to any area in which rocks (and their associated fossils) may be studied in situ.

How to use this book

The introductory chapters provide you with an understanding of how fossils are formed and of the variations that exist in terms of their preservation. You will also learn how best to find and collect your own material and relate both fossils and the enclosing sedimentary rock to an appropriate environment of deposition. The first of a series of keys (see page 38) will then help you decide on the group to which your fossil belongs, whilst the other keys which follow will help with a more precise identification. By using the keys you will also be able to turn to the correct colour plates and descriptions and confirm your identification. The illustrated glossary will help define some of the words which crop up regularly in the text.

Conservation and a code of conduct

Geologists and landowners are aware of the need to protect important localities. We must prevent vandalism and over-collection. The Geologists Association of Great Britain has drawn up a detailed code of conduct and this has been published in several countries. The main points to observe are:
1 Always obtain permission from the landowner before venturing on to private land;
2 Do not hammer at outcrops aimlessly, and never leave loose rock fragments scattered over fields and roads;
3 Close all gates;
4 Follow any local bylaws;
5 Remember that whilst your own safety is your responsibility, irresponsible behaviour may put others at risk.

Field equipment

You should always wear suitable warm and waterproof clothing and stout footwear, and carry topographic maps of either 1:50 000 or 1:25 000 scale. No geologist should travel without a notebook. Detailed fieldnotes are an essential part of your own records and may in time be the only reference to an outcrop destroyed by the spread of urban areas or erosion. Make sure your notes are clearly written and expressed, and that good sketches accompany your description. Photographs will help record both the overall geometry of the outcrop and small-scale features. They are, however, no substitute for a detailed sketch. All photographs should be numbered and captioned appropriately. Sharpened hard lead pencils, permanent ink 'rotring' pens, a rubber and coloured pencils are essential to the maintenance of a neat and tidy notebook.

Other pieces of equipment will include a hand-lens, sieve, penknife, safety goggles, compass clinometer, hammer and chisels. Of these, the hand-lens is most important, as it enables you to recognize fine details of both rock and fossil. The compass clinometer is vital should you wish to record details of outcrop geometry or take compass bearings for location purposes. In the field, keep your map on a clipboard and place both inside a plastic bag. A stout rucksack or canvas bag will be needed to carry these and the remainder of your equipment.

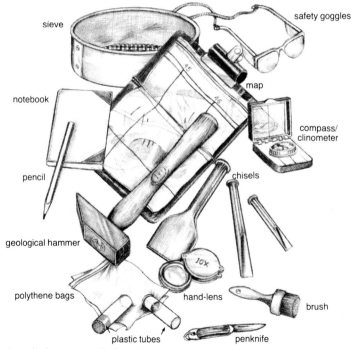

A notebook, pens, pencils, hand-lens, hammer, chisels and compass/clinometer are among the major items of equipment carried by a field geologist. Other less vital items are also shown.

Fossils – the remains of a past life

When a plant or animal dies its remains are often ravaged and destroyed by the weather, or eaten by animals in search of food. This is particularly true of land-living organisms where the bacterial processes of decay also help in the destruction and ultimate disappearance of both the soft tissues and the hard skeleton. Occasionally, an animal or plant will fall into water and sink into the soft muds present on the bottom of a pond or lake. There the remains will be buried and protected and, under the right conditions, the agents of decay will be unable to function. With time, the addition of more mud layers will give further protection, and chemicals within the surrounding waters may aid in the preservation of both the soft tissues and the skeleton. Such remains are potential fossils.

The remains of mammals, including human beings, are among those found 'pickled' in peat bogs – another form of fossilization. Their detailed preservation is outstanding, with hair, skin and internal organs providing clues to age and diet. These remains are very young in geological terms, however. The two or three thousand years which have elapsed since the animals were first buried in the bogs represent only a millisecond on the geological clock. The further we go back in time the less likely we are to find perfectly preserved organisms. The nature and character of these organisms will also vary as different groups of plants and animals were predominant at different periods of time and in different environments. Their modes of preservation will vary in response to the differing processes involved in the conversion of soft sediments to sedimentary rocks, and as a result of compaction and changes in chemical composition. Fossils are therefore the remains of plants and animals preserved within the sedimentary rocks. The perfectly preserved specimen is a rarity; the most common fossils are the hard shells and skeletons of animals, mineralized fragments of wood, and the tracks, trails and burrows formed by animals in soft sediment.

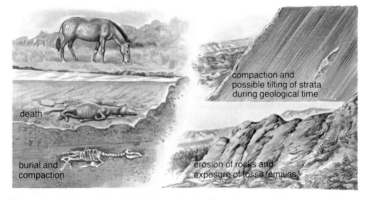

compaction and
possible tilting of strata
during geological time

death

burial and
compaction

erosion of rocks and
exposure of fossil remains

When a fossil dies, rapid burial in fine-grained sediment is essential for its preservation.

Modes of preservation

In most cases the process of **fossilization** is dependent on rapid burial in a protective medium and the prevention of decay. Bacterial action is limited or prevented by burial in muds or under volcanic ash, or by low temperatures. Very dry conditions can also result in the preservation of mummified organic remains. As already stated, the vast majority of fossils are of organisms which possessed hard parts, and the remains of marine animals far outnumber those of land dwellers. The animal kingdom is divided into animals with, or without, backbones. The former are called **vertebrates**, the latter **invertebrates**. Of the two the majority of vertebrates have skeletons composed of calcium phosphate while invertebrate shells and skeletal frameworks may be built of calcium carbonate, calcium phosphate, silica and various complex organic compounds.

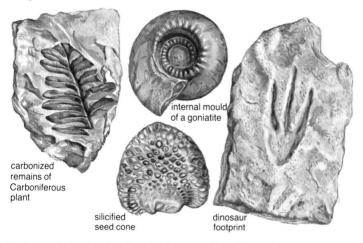

carbonized remains of Carboniferous plant

internal mould of a goniatite

silicified seed cone

dinosaur footprint

Fossils may be found unaltered, replaced or as moulds and impressions.

UNALTERED HARD PARTS   Among invertebrate animals the shells of clams and the hard, segmented bodies of insects are frequently found unchanged in rocks less than 60 million years old; in these instances the original structure of the hard parts, and their chemical composition, bear comparison with living relatives. Clam shells are composed of calcium carbonate coated with a thin, black organic layer. The calcium carbonate may occur in two forms: namely with calcite as the hexagonal-rhombohedral form, and aragonite as the orthorhombic form. Aragonite is easily altered to the more stable calcite, and thus aragonitic shells are rare in rocks older than 60 million years old.

Hard parts composed of calcium phosphate are resistant to chemical change, and the remains of various animals may be found unaltered in rocks 600 million years old. The same is true of skeletons made of complex organic compounds – the 'horny' cuticles of insects and their relatives being worthy of note. Relatively few organisms secrete the substance known as silica as their primary skeletal substance, but it is, however, common among groups of sponges. The original opaline silica of sponges is relatively unstable, however, and as with aragonite is rarely found in ancient rocks.

ALTERED HARD PARTS   The processes involved in the formation of a sedimentary rock frequently affect the fossils enclosed within the sediment. The continued accumulation of material exerts pressures on the buried deposits, and results in the expulsion of water and the compaction of grains. The fossils themselves therefore may also be distorted physically. The fossils may also be affected by changes in the water chemistry within the rocks.

The shells of various invertebrate animals are made up of minerals possessing a distinctive fibrous or layered structure. The retention of this structure indicates that the hard parts are probably unaltered. A mosaic or granular interlocking texture (which can be observed with a powerful hand-lens or microscope) will suggest **recrystallization**. This process is the result of the original fibrous mineral going into solution and re-forming with a coarser growth structure. The recrystallization of aragonite into the more stable form calcite will result in the loss of the original texture. Recrystallization does not commonly involve a change of mineral type, however.

Replacement, on the other hand, infers that the original mineral of the fossil is replaced by another. The process involves the percolation of fluids through the parent rock. The fluids react with the enclosed fossil and the original mineral passes into solution; almost simultaneously a different mineral is precipitated to fill the space left by the fossil. A common replacement mineral is pyrite – the golden, box-like crystals of which often enhance the appearance of the fossils. Pyritized fossils are frequently found in black mudstones (fine-grained rocks deposited in environments in which oxygen in lacking).

Silica is another common replacement mineral. Quartz or chalcedony will often replace calcite in shells or in the skeletons of corals. In the field this may be observed when such fossils stand 'proud' of a limestone surface. What has happened is that water rich in carbon dioxide has acted as a dilute acid and has etched the limestone surrounding the silicified fossils. Silica is also known to impregnate fossil wood, and here the delicate tissues are petrified or literally made into stone. This process is called **petrification**, and the fossil is of course made denser and heavier by the process.

Wood and plant materials in general are rich in volatile substances. The same is true of some invertebrate skeletons composed of organic materials, and both can be affected by the process of **carbonization**. This results in a decrease in the original volatile substances such as oxygen, hydrogen and nitrogen, and in the preservation of the leaf or skeleton as a thin film of carbon.

In permeable (porous) rocks, where water can pass along fractures and through unfilled spaces between constituent grains or particles, the fossils may be dissolved out to leave a series of cavities. This is a common phenomenon in sandstones, sandy limestones or ironstones, the cavities being termed **external moulds**. On occasion you may find that the original shell had previously been filled with sediment, and as a result of the shell dissolving the core or sediment now rests loosely in the mould. Such cores are termed **internal moulds** or **steinkerns**. In the case of clams the steinkern will reflect the internal form of the animal, and you may be able to recognize the size and position of the muscle scars and perhaps the area occupied by the internal organs. If the empty external mould is subsequently filled by sediment or mineral deposits, the new material forms a **cast** of the original shell.

It is possible to make moulds and casts from rocks in which the fossils have been dissolved. You can use plasticine or better still a cold-set rubber, which you simply pour over the rock surface. When dry, peel back the rubber carefully, making sure delicate traces are not torn from your replica. Fine details of the original shells will often be preserved on the surfaces of the cavities and these will now ornament a three-dimensional cast. Such casts provide additional data on the size and shape of the original fossil.

Pyrite is a common replacement mineral. It often enhances the preservation of fossils such as brachiopods (shown here) and ammonites.

Types of fossils

Most of the fossils you will find represent the hard parts of an organism, such as a shell, tooth or bone. Others, however, including casts, moulds and **impressions**, are of equal importance in providing details of individual organisms or fossil communities. Shells and skeletons are frequently termed **body fossils**, as they provide details on the shape and functions of the actual organism. The same may be said for an impression – particularly when it retains details of the soft parts. Body fossils occur in all shapes and sizes, ranging from microscopic sea-dwellers to huge terrestrial dinosaurs. Their preservation will vary according to the conditions that prevailed at the time of death and burial.

SUBFOSSIL This is a term sometimes applied to the remains of animals and plants preserved in rocks less than 10 000 years old. These include the remains of bison trapped in peat bogs, or of ancient man mummified in caves. Subfossils were formed after the last ice age, during the Holocene epoch.

MICROFOSSILS These are usually less than 0.5 mm ($\frac{1}{50}$ in) in size, but organisms usually regarded as microscopic can deposit skeletons up to 10 cm (4 in) in diameter. Both single-celled plants and animals can form mineralized skeletons, and some make a major contribution to the formation of sedimentary rocks.

Animals such as the woolly mammoth that lived during the Pleistocene and Holocene are sometimes found frozen in peat bogs. They are best described as subfossils.

MACROFOSSILS   These are greater than 1 cm ($\frac{2}{5}$ in) in size. The term is usually applied to the more advanced plants and animals, such as clams, corals or the skeletons of vertebrates.

UNUSUAL FOSSILS   These are, by definition, extremely rare. They include mammoths dug from the Siberian wastes and the remains of the first bird *Archaeopteryx*. The term 'unusual' refers to the mode of preservation, in which a combination of events and conditions results in all or most of the organism being preserved in the rock. Famous deposits include the Solnhofen Limestone of southern Germany and the Burgess Shales of Canada.

TRACE FOSSILS   These are formed by organisms performing the functions of everyday life, such as walking, crawling, burrowing, boring or simply feeding. Dinosaur footprints, worm trails and clam burrows are all trace fossils.

COPROLITES   These are also trace fossils. They are the droppings of animals. They can vary in size from the tiny **faecal pellets** of a sea-snail to the large coprolites of crocodiles, dinosaurs or mammals.

BIOCLAST   This is the term given to fossils or fragments of fossils enclosed in sediments. It is usually applied to a hand specimen or to a thin section under the microscope.

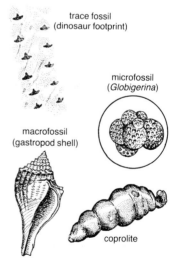

trace fossil
(dinosaur footprint)

microfossil
(*Globigerina*)

macrofossil
(gastropod shell)

coprolite

Archaeopteryx is the earliest recorded bird. Its remains are found in the Jurassic Lithographic Limestone of southern Germany.

The fossils illustrated above represent examples of the groups described on pages 12 and 13.

The distribution of fossils in time and space

Geologically speaking, time and space are difficult to determine accurately without the collection of a vast amount of data on sedimentary sequences and the fossils they contain. Sequences of rock which share specific characteristics such as grain size and/or sedimentary structures are known as facies. Some fossils may be restricted to given facies, others to a specific horizon or time zone. They may be defined as follows.

FACIES FOSSILS These are found in association with a particular sequence of sediments which have given characteristics. Facies fossils are of little or no use to the stratigrapher, because they may be limited in terms of geographical distribution. However, they may be useful in the interpretation of environments.

ZONE FOSSILS These are restricted to specific levels within the stratigraphic column. These levels are called biostratigraphic units, and may be defined on the range of a given fossil or in the case of an **assemblage zone** on total fossil content. The term is synonymous with **index fossil**. The most useful zone fossils are short-lived and widespread in their distribution.

RANGE This is used to describe the time interval during which a fossil existed on Earth. This time interval is measured in terms of rock units which may be part of a period, or may possibly span several periods, of geological time.

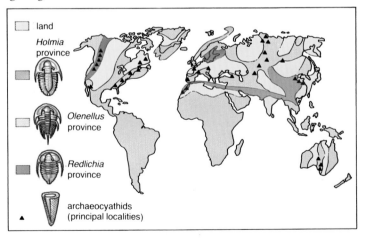

An area characterized by a group of fossils and defined by either geographical, ecological or climatic factors is referred to as a faunal province.

DERIVED FOSSILS   These are remains that have been eroded from one bed, and then transported and deposited in a younger (more recently formed) bed. They are therefore older than the sediment in which they are enclosed when discovered.

FAUNAL PROVINCES   These define the distribution of associated groups of organisms. The boundary of the province may be marked by climatic or geographic criteria. In the stratigraphic record (*see* page 17) it is almost impossible to recognize these criteria with any degree of accuracy. A province is therefore identified on the association of fossils in one area at a particular interval in the geological time-scale. The association may be referred to as a **faunal assemblage**. Faunal provinces may vary in terms of their spatial distribution. An accurate reconstruction of a faunal province will probably provide information on the Earth's plate movements and the palaeogeography of the period.

LIFE ASSEMBLAGE   This is used to describe a group of organisms that have remained in situ after death. Reefs are good examples of life assemblages.

DERIVED ASSEMBLAGES   These are accumulations of organisms that have been transported and subsequently buried outside their normal habitat or living area.

The term life assemblage is used to describe a group of fossils that were buried in 'life' position.

Death assemblages are essentially accumulations of fossil material deposited by water outside the immediate area where they lived.

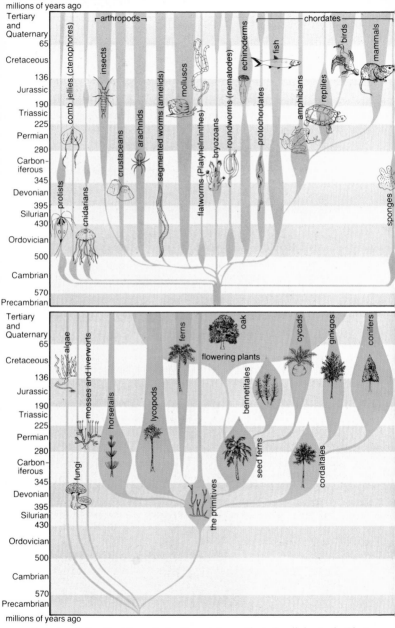

Animal evolution (top) began in the Precambrian, with single-celled animals. Plant evolution (above) also began in the Precambrian.

Geological time-scale

Our planet is approximately 4.6 billion years old. This age is established by using dating techniques – called radiometric dating – which measure the rates of radioactive decay in given rocks. Of the complete span of geological time the **Precambrian** lasted slightly in excess of 4 billion years. Precambrian rocks are commonplace in **continental shield** areas, such as the Canadian Shield. They are mostly altered or **metamorphosed** by heat or pressure. Fossils are rare in these rocks; the first communities appearing about 670 million years ago. The Precambrian is divided up according to the types of rock and their textures. The 600 million years from the Precambrian to the present day are marked by the presence of various forms of life. The term **Phanerozoic** meaning 'obvious life' is applied to this span of geological time. The superimposition of sedimentary layers is another feature of this timespan, and the evolution and extinction of many organisms provide a key to the subdivision of the Phanerozoic. Subdivision by the grouping of layered rock sequences is termed **lithostratigraphy**, whereas grouping by the use of fossils is known as **biostratigraphy.**

Another branch of **stratigraphy**, the study of stratified rocks, is called **chronostratigraphy**. It examines the distribution and correlation of rock types throughout the world at a specific period of time. Radiometric dating helps establish a more precise age.

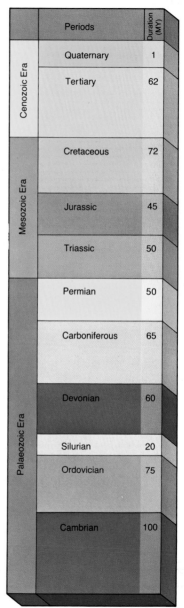

Periods	Duration (MY)
Cenozoic Era	
Quaternary	1
Tertiary	62
Mesozoic Era	
Cretaceous	72
Jurassic	45
Triassic	50
Palaeozoic Era	
Permian	50
Carboniferous	65
Devonian	60
Silurian	20
Ordovician	75
Cambrian	100

The geological time-scale is divided into 11 major periods.

Evolution

Evolution is the process by which new forms of life develop from existing organisms. The fossils described and illustrated in this book are a testimony to the many thousands of minute changes that have taken place over millions of years. Specific periods of geological time are noted by the presence of given groups of animals and plants, and in many cases it is possible to identify ancestor/descendant relationships among these animals and plants.

Starting in the Precambrian, the variety of organisms is initially restricted to single-celled plants and bacteria. Approximately 2000 million years ago the first single-celled animal appeared, with the earliest multicellular animals appearing around 1000 million years later. Soft-bodied animals such as jellyfish and simple hydrozoans (relatives of the sea-anemones) were common worldwide 700 million years ago. Animals with hard parts mark the beginning of the Cambrian Period, and gradually the various groups of invertebrates, vertebrates and higher plants are added to the fossil record. Their appearances are an indication of an increase in overall complexity, with man (*Homo sapiens*) topping the evolutionary pyramid. The appearance of man is the culmination of all the minute evolutionary changes that have gone before.

The palaeontologist can study changes in the shape and form of related fossils. He can recognize the presence of primitive characters within a shell or skeleton, and can record the new features. Only animals and

Eocene	Oligocene	Miocene
Hyracotherium	*Mesohippus*	*Merychippus*

plants best fitted will survive, with a given character or group of characters enabling a specific animal or plant to flourish, whilst its contemporaries become extinct. Biologists know that the characteristics of any plant or animal are passed on to successive generations through units of inheritance, called **genes**. These are chemical codes which are important in the process of the natural manufacture of proteins. Variations in the proteins lead to variation within a species. In the fossil record the influence of these chemical changes is reflected in, for instance, the length or shape of a tooth or the number of chambers in a shell.

Through the collection of fossils we can recognize the course of evolution for a given species or group of species. We can also plot lines of descent and recognize the development of new stocks from an ancestral population. The evolution of the modern horse from the terrier-like *Eohippus* (*Hyracotherium*) of the Eocene is well documented. It involved changes in dentition, a lengthening of the head and neck, a strengthening of the leg and the fusion of the toes to form a hoof. Gradually the species increased in size, and the horse changed from a forest glade-dweller to a grass-eating, prairie-dwelling animal. In North America the evolution of the horse is recorded in the Eocene, Oligocene, Miocene and Pliocene sediments of Dakota, Nebraska and neighbouring states. In Europe and Asia the record is rather incomplete but remains of the first horse are known from the London Clay Formation of England.

The evolution of the horse can be charted through changes in the teeth (including crown patterns), foot bones and overall size.

	Pliocene	Pleistocene
	Pliohippus	*Equus*

Recognizing sedimentary rock types

Fossils are usually associated with **sedimentary rocks**. These rocks are produced by weathering and chemical processes. You can observe these processes at the foot of a steep mountain slope, on a beach, in hot climates or in lakes bordering seas and oceans. Ice will grind the surface of a rock to produce a powdery rock substance called **rock flour**, whereas frost will assist in the mechanical breakdown of the same material by freezing the water trapped in cracks and crevices, causing shattering. The resultant rock fragments will be angular and form screes in mountainous areas. A combination of intense radiation (due to the desert sun) and rapid cooling will also give rise to a fragmental rock. The action of waves pounding on a beach will result in rocks composed of fragments of particles, but in this case the particles will be more rounded. This may indicate the amount of transportation that has taken place, and geologists will refer to the level of maturity of a sandstone based on the degree of roundness of individual particles.

Rocks composed of rock, mineral and fossil particles are termed **clastic rocks**. They are subdivided according to grain size, with three main groups recognized by sedimentologists. These are **mudrocks**, **sandstones**, and the coarser grained **breccias** and **conglomerates**. **Non-clastic rocks** are those formed by the precipitation of minerals and/or the accumulation of plant and animal remains. They include many limestones and the evaporites. They are subdivided on their mode of origin, chemical composition and textures.

Clastic rocks

MUDROCKS   These are the finest grained of the clastic rocks, and individual grains are not visible to the naked eye. They are usually less than 0.0625 mm ($\frac{1}{400}$ in) in size and the main constituents are clay minerals, quartz, feldspar mica and calcite. These fined-grained rocks are also described as **argillaceous**.

Clays are the very finest of mudrocks, with grains usually less than 0.004 mm ($\frac{1}{5000}$ in) in size. Clays have a blocky fracture. Claystones are formed when clays are compacted and lose water. Muds and clays often yield rich faunas of molluscs, small corals and brachiopods.

Shales are compacted mudstones (also a variety of fine mudrock) in which the flat minerals, including clays and micas, are realigned along a preferred plane of orientation. This often gives the shale a degree of flakiness and a surface sheen or glitter. Should the amount of fine-grained quartz or feldspar increase, the shale will tend to form thicker, harder units. These may be referred to as tilestones or flagstones. Very thinly bedded shales that flake or peel are called paper shales.

Siltstones may contain quartz, feldspar mica and calcite. They have a low clay content and the grains are usually larger than 0.004 mm ($\frac{1}{5000}$ in) in size. Brachiopods and bivalves are among the more common fossils associated with these sediments.

You can test the calcareous content of shales or siltstones by pouring white vinegar over the rock. If it effervesces then you know that calcium carbonate is present. The term **marl** is often given to a mudstone with a calcareous cement. Geologists often use the term mudstone to describe a very fine-grained rock in which the grains are cemented together and which does not become plastic (soft and pliable) when wet.

Fine-grained sediments often indicate calm conditions of deposition. The presence of pyrite will also suggest the absence of oxygen and bacteria. A low level of water movement and few agents of decay will therefore give rise to well-preserved fossil material. This is often true of fine-grained rocks with black shales containing the remains of many organisms that floated or swam in open seas. Few sea-floor organisms could be expected to live under such inhospitable conditions, however. Graptolites, goniatites and thin-shelled bivalves are common fossils in Palaeozoic black shales.

SANDSTONES   Unlike the finer-grained mudrocks, the composition of various sandstones can be determined quite well with a hand-lens. Individual grains can vary from angular to well rounded, the degree of roundness indicating the degree of maturity. Sandstones consist mainly of three main components: quartz, feldspar and rock fragments. Of these, quartz is a harder, more stable mineral than feldspar and will survive more than one or two cycles of weathering, erosion and deposition. Feldspars will break down relatively quickly to give clays. Rock fragments are obviously derived from existing Earth materials and will gradually break down into their consistent minerals.

QUARTZ SANDSTONE   Sandstones consisting of well-rounded quartz grains are among the most mature sediments. If the matrix between the grains accounts for less than 15 per cent of the rock then the sandstone is termed an arenite. When the arenite is well cemented it is often referred to as a quartzite. If the matrix exceeds 15 per cent, then the rock is called a wacke, with quartzwackes being more mature than those with unstable minerals. Quartz sands are often porous and the fossils they contain may be dissolved to leave casts and moulds. Vertical burrows, wood fragments and shelly faunas are the most common fossils in such sandstones.

ARKOSE   This descriptive term is used when you can recognize the presence of feldspars, quartz and rock fragments within a sediment of sandstone grain size. Feldspars would account for 25 per cent or more of the rock, and rock fragments 50 per cent or less.

LITHIC SANDSTONES   These are sandstones in which rock fragments account for more than 50 per cent of the rock. They also have a very low matrix percentage. Other sandstones can be classified on the presence of a specific mineral. Glauconite, mica and phosphate can characterize a specific sandstone, and each indicates a particular environmental association. Muds and minerals such as iron, calcium carbonate and silica can cement the grains within a sandstone. The names argillaceous sandstone, ironstone, calcareous sandstone and siliceous sandstone are therefore used to describe such sediments.

Sandstone grains range in size from 0.0625 mm $(\frac{1}{400}$ in) to 2 mm $(\frac{1}{10}$ in). The colours of sandstone can be quite distinctive, but colour does not necessarily indicate environment. The colour is related to the chemistry of the cement and the constituent minerals. Red sandstones are often the result of the oxidation of iron, and the thick, red-stained successions of the Permo-Triassic are associated with terrestrial environments. Sandstones are formed on land, in rivers and deltas and in shallow seas. They often indicate relatively high-energy conditions with wind and water as related agents. Terrestrial sandstones may contain plant remains, whereas materials deposited from water will often retain evidence of both abundant trace and body fossils. Sandstones are classified under the heading of **arenaceous rocks** in most text books.

CONGLOMERATES AND BRECCIAS These rocks are coarse grained, with grains normally exceeding 2 mm $(\frac{1}{10}$ in). **Conglomerates** are recognized by the presence of subrounded to rounded clasts; **breccias** by angular ones. Once again, the degree of rounding is an indication of maturity. Conglomerates can be classified by the type of pebbles within them; they are often linked with stream, lake or sea shore environments. Because of their angular fragments breccias indicate a limited degree of transportation. Screes on mountain slopes are good examples of sedimentary breccias. Fossils are found infrequently in association with conglomerates and are even rarer in breccias. These coarse-grained rocks are also termed rudaceous.

Non–clastic rocks

This group includes a variety of rocks formed by the precipitation of minerals and plant remains. For palaeontologists the most important groups of non-clastic rocks are limestones and corals.

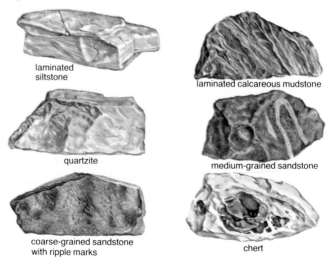

laminated
siltstone

laminated calcareous mudstone

quartzite

medium-grained sandstone

coarse-grained sandstone
with ripple marks

chert

LIMESTONES Limestones are rocks in which the proportion of calcium carbonate exceeds other constituents. Calcite is the most common mineral in limestones; it may occur as very fine crystals or in the skeletons and shells of invertebrate animals.

Of the three schemes currently used to classify limestones, the simplest to use in the field is one that combines the determination of the main grains or components with **matrix** type. (The use of this somewhat limited classification avoids the obvious mistakes one may make in trying to recognize fine detail without the benefit of examination of a thin section of the rock under a microscope. If the matrix is fine and rather mud-like, then it is termed **micritic**; the calcite component having a microcrystalline form. Micrite, or microcrystalline calcite, probably originates with the breakdown of the external skeleton of calcareous algae. Should the limestone have spaces filled with a coarser crystalline calcite, and have little or no micrite present, it is termed a sparite.

Should either micrite or sparite contain the remains of plants and animals, they are termed **biomicrites** or **biosparites**. Those containing rounded particles, formed by the gradual accumulation of carbonate minerals around organic or inorganic remains, are called **oomicrites** or **oosparites**. The rounded particles are termed **ooliths**. The degree of fragmentation of organic remains is an indication of the energy level at the time of deposition. Broken fragments will suggest a higher level than it would for whole specimens. The organic fragments are known as **bioclasts**.

Microfossils belonging to the Foraminifera are common components of both fine- and coarse-grained carbonate rocks. The smallest fossils were planktonic in habit. Their tiny shells settle on the sea floor and

Sediments are subdivided according to grain size, mineral composition and texture.

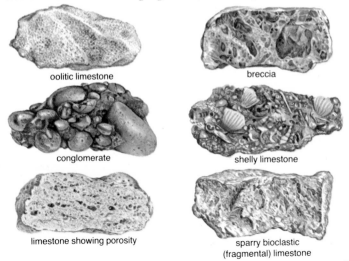

oolitic limestone

breccia

conglomerate

shelly limestone

limestone showing porosity

sparry bioclastic
(fragmental) limestone

accumulate to form an ooze. Other microscopic organisms, including plants such as the diatoms and coccolithophores, also form deep-water oozes. The best known of these is the famous white chalk of northern Europe.

Shells of the larger Foraminifera are also major contributors to the formation of limestones. Different families dominate specific episodes of geological time, with the many chambered shells of the coin-like *Nummulites* characteristic of Tertiary Mediterranean limestones. The larger Foraminifera are mostly indicators of shallower water environments.

Algae, including the coccolithophores noted above, have been important 'rock-formers' for billions of years. **Stromatolites** – layered, domed structures common in the Precambrian and early Palaeozoic – were formed when algae trapped grains on their sticky surfaces. In the shallow waters and on beaches the grains may be arranged concentrically or form asymmetric structures called **oncolites**. Both oncolites and stromatolites are large-scale elements, as are the frameworks formed by calcareous algae in reef environments. Other algae may disintegrate due to wave action and provide a high input of micrite particles into the marine environment.

Corals are major constituents of limestones during the Palaeozoic, Mesozoic and Cainozoic. They are often associated with algae and bryozoans as reef builders, with many other organisms encrusting or boring into the fabric of the build-up. Reef-like mounds may also be formed by **rudist bivalves** and **archaeocyathids**. Accumulations of gastropods, bivalves and brachiopods are common in fossil record. The term shelly limestone is used to describe limestones rich in such fossils or their debris.

Accumulations of shells or shell debris are also referred to as **coquinas** or **lumachelles**. Brachiopods are common in limestones of the Palaeozoic, whereas bivalves and gastropods are more important in those of the Mesozoic and Cainozoic. Other organisms involved in the trapping or accumulation of carbonate sediments include worms and echinoderms.

EVAPORITES/PHOSPHATE ROCKS AND NODULES   Evaporites are mainly chemical rocks formed when dissolved salts, concentrated by evaporation, precipitate out as massive or nodular deposits. **Rock salt** or **halite** is a typical evaporite formed from the evaporation of saltwater lakes, lagoons or shallow seas. Gypsum often grows as nodules in mudrock sequences and is associated with **sabkha** (supratidal) environments. Algae and gastropods are the most common organisms associated with these types of deposit.

Most phosphate rock is formed on areas bordering the edge of the continental shelf. The rock is often nodular or pelleted, with organic materials often replaced by phosphate minerals. The areas of deposition were rich in microscopic plant materials and as a result fish and large fish-eating animals are common. Phosphates are frequently rich in vertebrate remains and trace fossils.

Sedimentary environments

It is often said that the best geologist is the one who has seen the most rocks. This is only partly true, but an in-depth knowledge of various sedimentary rocks, their textures, associated structures and fossils is of great importance in the interpretation of the environment in which the sediment was deposited. The degree of roundness, the size of individual grains, the variety of minerals and rock fragments and the type of matrix all provide information on the rock itself. Sedimentary structures, formed when the rocks were laid down, reflect the behaviour of individual grains and the way they were transported prior to deposition. In normal sequences of sedimentary rocks you will be able to identify beds of varying thickness. These are laid down on the bottom of a lake or on the sea floor. Each bed marks the imput of sediment over a period of time; it is usually marked both at the base and at the top by a distinct horizontal plane. This marks a surface of deposition and is termed a **bedding plane**.

Some bedding planes may be marked by the presence of abundant body and trace fossils. Not all of the planes within sedimentary rocks are horizontal. When the ripples in sands or carbonate sands migrate, individual grains move up the shallow slope of the ripple and down into

Sedimentary structures are formed during deposition, with specific structures reflecting the response of suspended particles to fluctuations in current strength and direction.

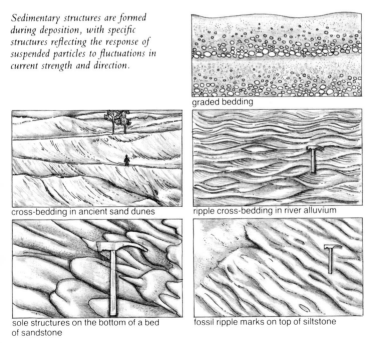

graded bedding

cross-bedding in ancient sand dunes

ripple cross-bedding in river alluvium

sole structures on the bottom of a bed of sandstone

fossil ripple marks on top of siltstone

the trough. The same is true for dunes on land or in shallow waters. In section the ripple or dune will show an internal structure in which the layers are at an angle to the horizontal. This is termed **cross-stratification** or **cross-bedding**, but is sometimes referred to as **false-bedding**. The scale and geometry of the cross-stratification is important in the determination of the energy level and direction of transportation. An alternation of ripple-marked sands and thin mud drapes are often a good indication of tidal flats. The large-scale cross-stratification of the Neo Red Sandstone of Devon and south-west England is the result of **aeolian** activity; wind-blown sands forming distinct dunes typical of a desert region.

Shrinkage cracks and **rain prints** in fine-grained sediments are indicative of exposure and desiccation. Another indication of rocks exposed to weathering is the potholed and cavernous surfaces found on limestone pavements, such as in the Peak District in Derbyshire, England. In the stratigraphic record these surfaces will represent a break in deposition. They are called **palaeokarst surfaces**. The plane representing the break in the succession is termed an **unconformity**.

Terrestrial and non-marine environments

Sand dunes, shrinkage cracks, rain prints and palaeokarst surfaces are clues to terrestrial environments. Mostly these are environments in which clastic rocks are formed. Silts, sands, conglomerates and breccias are the main rock types.

DESERT ENVIRONMENTS These are associated with subtropical, arid regions. Sandstones, conglomerates associated with flash floods, and silts associated with temporary playa lakes, are the diagnostic sediments. Fossils are rare but include dinosaurs from the Cretaceous deposits of North America and fossil forests and dinosaurs from North Africa.

RIVER ENVIRONMENTS Sediments associated with these environments vary from mudrocks to coarse conglomerates. The rock types that occur in different environments are summarized diagrammatically on pages 28 and 29. Plant remains and trace fossils may be linked with certain river environments, but vertebrate and invertebrate remains are usually rare.

DELTAIC ENVIRONMENTS These are found at the mouths of rivers. They are characterized by the thick accumulation of sediments. This is because the flow of the river is slowed rapidly and the suspended materials settle to the lake or sea floor. Deltas have birdfoot or lobate shapes, and cross-bedding is a common feature. The areas bordering deltas are often densely colonized by plants, and plant debris is a common feature of delta sands and silts. The deltaic deposits of the Carboniferous coal fields are renowned for the plant fossils – these are usually associated with non-marine clams and marine cephalopods, and indicate that fresh and seawater conditions fluctuated over the delta and its environs.

LAKE AND SWAMP ENVIRONMENTS These are often associated with delta plains. They can also occur in the lowlands bordering seas or in faulted areas such as rift valleys. Lake or lacustrine sediments will be characterized by a mixture of sandstones and fine silts and muds. These sediments form distinct sequences which, like those of deserts or deltas, may be termed facies. The groups of sediments or facies associated with lakes may contain invertebrate, vertebrate and plant fossils. In the lake sediments of the Kenyan Rift, the remains of many mammals, including those of early man, provide clues to our own evolution. Swamps or paludal sediments are often finer grained, and beautifully preserved fossils are a common occurrence.

Marine environments

If one drew an imaginary line from the sand dunes bordering the beachline seaward to the middle of an ocean, it would traverse a variety of environments. These would include high-energy beach areas, lagoons, the continental shelf and the oceanic basin. Once again sedimentary sequences or facies provide a distinctive clue to the type of environment.

Areas above high tide vary from high cliffs to sand dunes and coastal swamps. These are the areas beyond the beach and are termed supralittoral. Whether wind-blown sands or hard rocks, they may still retain evidence that when fossilized would provide clues to the reconstruction of the environment. Usually this is in the form of trace fossils with vertical burrows or borings, indicating the animals' need for protection and shelter.

BEACH ENVIRONMENTS Beaches are often composed of linear bands of sand or pebbles. These represent zones of high wave activity, and the sediment grains are mostly rounded. The mineralogy of the beach sands will vary in relation to the materials being eroded. In the Bahamas or along the Kenya coast the sands are calcareous and the grains consist of shell fragments. The term **littoral** is applied to the zone between high and low tide. In many areas the beach will be bare of sediment, the country rock itself providing a hard substrate. In these areas seaweeds will flourish together with various invertebrates that attach themselves to the surface or bore shallow dwellings. In soft, mobile sands, plants find difficulty in fixing themselves securely, whereas the dominant animals are burrowers. In the stratigraphic record the bored surface of the carboniferous limestone at Vallis Vale in the Mendips of England is evidence of the movement of the Jurassic sea over an exposed hard rock surface. At Thionville sur Opton in the Paris Basin, France, a sequence of beach gravels and sands contains abundant burrowers.

LAGOONS AND RESTRICTED BAYS These are found behind reefs or sand shoals and are, by comparison with the beach or reef, low energy environments. Sediments again reflect the source materials but are usually fine-grained sands or muds. In warm tropical areas the sediment will owe much to the destruction of algal remains. Elsewhere the muds may be composed entirely of faecal pellets. The Upper Jurassic limestones of the Dorset coast, England and the Lowville Limestone in New York, USA, are examples of lagoonal limestones.

SHELF ENVIRONMENTS

The continental shelf stretches from beach to shelf edge, which is usually defined as 320 km (200 miles) from land. The shelf may slope gently into the oceanic depths or may fall suddenly into a deep basin. A gently dipping shelf is termed a **ramp**; the steeper type a **drop off**. The facies associated with shelf environments may be divided into broad zones, the character of which is determined by the interplay of various factors. Wave energy will obviously have an effect on sediment grain size, whilst the amount of light, temperature, food, depth of water, turbulence and type of sediment will control the distribution of organisms.

Sandstones, fragmental limestones and calc-arenites will be typical of the shallower shelf areas. These may be rich in fossils, with clams, gastropods, brachiopods, sea-urchins and oblique and horizontal trace fossils as the most common forms.

REEFS AND CARBONATE BUILD-UPS

These are often associated with warm, shallow shelf environments. Reefs are composed of various life forms, with algae, corals and bryozoans as the main framework builders. In the geological record archaeocyathids, stromatoporoids and rudist bivalves have played significant roles in the building of reefs worldwide. Apart from the main framework components, a range of organisms including worms, brachiopods, clams, echinoderms and crustaceans will be found attached to, or embedded within, the reef structure. It is possible to identify the back, top or front of a reef by examining the association between the plants and animals and the sediments which enclose them.

BACK REEF   slopes are littered with reef debris and interfinger shoreward

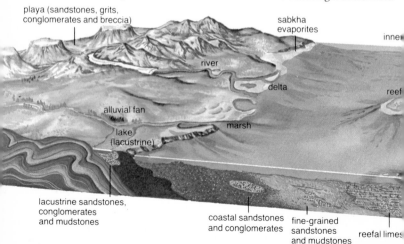

playa (sandstones, grits, conglomerates and breccia)

sabkha evaporites

inne[r]

river

delta

reef

alluvial fan

lake (lacustrine)

marsh

lacustrine sandstones, conglomerates and mudstones

coastal sandstones and conglomerates

fine-grained sandstones and mudstones

reefal limes[tone]

with lagoonal mudstones packed with faecal pellets and foraminiferans.
REEF FLATS OR TOPS   often consist of massive calcareous encrustations
on the upper surface supporting an abundance of fixed mobile organisms.
Seaward, the reef may be grooved or channelled.
REEF FRONTS OR FORE REEF   slopes are steep – often vertical. An array of
corals, algae and bryozoans flourish in the warm, well-oxygenated
waters of modern reefs. Fish and a variety of invertebrates thrive in this
zone. A slope of debris derived from the reef (known as a **talus slope**)
occurs below the reef front. Fine examples of fossil reefs are found in the
Silurian limestones of the Welsh borders in Britain and in the El Capitan,
Permian limestones of the Guadalupe Range in New Mexico.

Beyond the reef and inner shelf environments carbonate rocks may
persist for many kilometres. Carbonate build-ups or limestone
accumulations may be found also in deep-water shelf environments.

CONTINENTAL MARGINS   These may be characterized by
accumulations of either clastic or carbonate materials. Earth tremors may
generate slides or slumps of these materials down the continental slope.
The sediment masses may vary in terms of sorting and overall form.
Turbidites are gravity flow deposits that exhibit good bedding and
sorting, whilst grain flow and debris flow deposits are more brecciated
and lack grading of particles. Derived fossils may be present in all three
types of sediment. Trace fossils are common in many turbidites
sequences including the Aberystwyth Grits of western Wales.

BASIN ENVIRONMENTS   Continental shelf material may slump
over or mix with the fine-grained sediments of the ocean basins. Basinal
or deep sea sediments are rarely rich in sand grains and are usually muds,
clays or oozes. The latter are mainly composed of a preponderance of one
type of micro-organism, namely diatoms (algae), globigerinids or radio-
larians (foraminiferans). Once again trace fossils are common to certain

*The sedimentary environments illustrated are characterized by their own diagnostic
sequences of sedimentary rock types and sedimentary structures.*

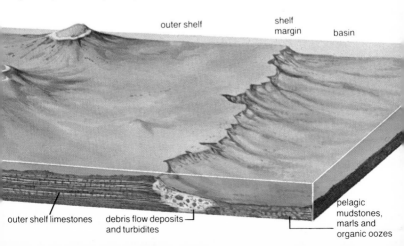

outer shelf   shelf margin   basin

outer shelf limestones   debris flow deposits and turbidites   pelagic mudstones, marls and organic oozes oxygen-rich deep-sea deposits. Deep-sea body fossils are often delicate and pyritized. The type of fossil varies with the type of sediment. In the Cretaceous of northern Europe the chalk is an example of a pelagic sediment. In North America the Lower Palaeozoic black shales, muddy limestones and cherts of the Appalachians are further examples of basinal deposits.

Whilst the deposits of lagoons, the shelf or basin are essentially specific to those environments, some deposits may be indicative of a single short-lived event. These include storms or tempests, and a relatively thin deposit may spread across a number of environmental areas. **Tempestites** are an example of these short-lived phenomena.

Cretaceous chalk of southern England.

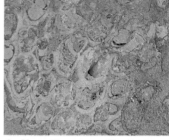

An Upper Cretaceous reef in Tunisia.

Fossils dunes indicate an ancient desert.

Raised beach left by drop in sea level.

Left: Sand dunes such as those found near Kerkaz in southern Algeria are representative of arid environments.

Collecting fossils

In order to get the most out of your fieldwork it is worth planning ahead. One of the first items to acquire is an accurate map of the area; 1:50 000 or 1:25 000 scale maps are the normal choice. Descriptions of the area, the outcrops and the fossils found there may exist in scientific journals, and these should be consulted if the best results are to be obtained. They will often contain grid references and so pinpoint the locality with great accuracy. Should your trip take you to the coast take care to consult tide tables. The equipment required for a successful fieldtrip is described on page 7.

Finding fossil-rich beds

Quarries, rocky outcrops and rocky coastlines are usually clearly defined on both geological and topographical maps. Many geological maps also indicate whether or not the rocks in these areas are fossiliferous. With this information you can plan ahead. Once you have decided on the general area you are going to visit in search of fossils, how will you know exactly where to begin looking for material when you reach your destination? This can be a problem in some areas, but usually there are clues to help you. Often you will notice that the rocks are layered, and dip in certain directions, or are horizontal. Differences in the colour and texture of the rock may tell you that different beds alternate with each other – the boundaries betwen them representing the bedding planes. If exposed as a flat surface a fossil-rich bedding plane will invariably yield the best materials and data. Specimens will be etched out over the surface, and in some cases will be easy to collect.

Vertical bedding planes like those in this quarry frequently offer excellent opportunities for collecting. Beware falling rock, however.

If the bedding planes are difficult to reach then the presence of fossil material may still be observed along fractured or jointed surfaces. It is usually difficult to collect fossils exposed in this way, as they are often embedded in protected, unweathered surfaces. You can, however, use this evidence to trace the bed until a weathered surface or bedding plane is exposed. You may also use features such as colour, type of weathering and bed thickness to trace a fossiliferous horizon along the strike of the bed or between outcrops. In limestone areas fracturing along joints, due to ice, may result in an accumulation of fossiliferous sediments on a scree slope (*see* page 22). This is true for other well-cemented and well-jointed sediments that occur as outcrops in mountainous areas.

On the seashore soft rocks will often collapse and flow, and finding a bedding plane is a difficult task. Often the fossils are buried in very thick sediments, and their discovery is mostly a matter of luck. At some localities it is therefore better to search along the foreshore as the fossils washed out of the cliff face tend to accumulate near the high-tide mark. Strandlines are often excellent sites for fossil collecting therefore. It is also possible that a fossiliferous horizon will be exposed on the beach, and a search over the flat rock pavement which may extend outwards along the beach will be more rewarding than using a hammer and chisel on a vertical cliff exposure. Even in areas where the rocks are vertical and the outcrop washed and rounded a 'rock grain' will be apparent in most sediments. Look for laminations, as well as colour banding and the other tell-tale features. These will often be related to bedding, and careful probing will reveal well preserved material. If you cannot find the bedding plane and the rock is not well jointed, it may be a waste of time trying to collect from that particular outcrop, but remember to record what you can of the outcrop in your notebook and on film anyway, using the techniques described below.

Fossils can be used to correlate rocks between outcrops, and to give the sediments in which they are enclosed a relative age.

Recording details

First, stand back and observe the exposure as a complete entity. Then draw a sketch and describe briefly the major features. (Fine-scale features can be entered on to your sketch or enlarged diagram later.) If the rocks are well bedded or exhibit cross-stratification then measurement of the **dip** and **strike** should be recorded. The dip is the amount by which the bedding plane (*see* below) is inclined to the horizontal. The strike is drawn at right angles to the dip. (A simple analogy is a roof, the ridge representing the strike and the roof slope the dip angle.) One method often used in the field is to pour water gently on to the inclined surface: the line it follows will normally represent the dip direction. In order to take **bearings** of your location you should sight the compass across to the first of several prominent landmarks, align the compass to north and take a reading. Repeat this exercise for subsequent readings from other landmarks approximately 90 degrees away from the first one. Other field workers will then be able to use your bearings and, by adding 180 degrees to each, draw lines from the landmarks to locate the position with accuracy.

Once you have recorded your position accurately and taken measurements, study the relationships of the individual beds, or layers of strata.

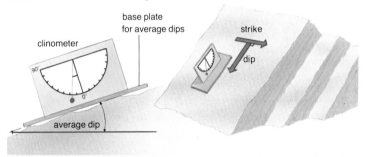

The greatest angle to be recorded on a smooth bedding plane is the angle of true dip. The direction of strike is at right angles to the angle of dip.

Using your compass, line up on a given landmark, read off the bearings against the midline of the compass, then repeat this exercise against another landmark approximately 90° away from the first to fix the position of your outcrop.

They may be described in terms of their thickness and the contact they make with one another. Thin, medium, and massively bedded are terms used in the description of a bedded sequence. Sketches will again help record the interesting features you observe. Once you have completed this task you can then describe the rock and fossils it contains. Remember to use descriptive terms in a scientific manner: for instance, 'fine-grained, buff-coloured limestone, containing rounded rock fragments and rare, small bivalves'. A fresh piece of rock, carefully removed by using your hammer, will provide this sort of information. The use of a hand-lens is essential.

After these important tasks are done, others related directly to the collection of fossils remain. For example, if the rocks are rich in fossils, sketches and photographs of their position and relationships should be made. Accurate measurements of the size and shape of individual fossils should also be entered into your notebook. Faunal communities can be recorded in terms of the numbers and positions of species; do this by counting and measuring the species within a one-metre square placed at regular intervals along a given line on the bedding plane. This data will be invaluable to others and will provide you with an in-depth knowledge of the specific horizon and of the time interval studied.

Extracting material

The irresponsible destruction of an outcrop in search of one or two fossils is not welcomed. Fossils should be collected sparingly, preferably without hammering. Often the best fossils are those which have been weathered out over a long period. These rest on the rock surface or among the scree at the foot of an outcrop. Collecting is a matter of judgement; if you think that the removal of the specimen is essential then do it, but first consider its scientific value and whether or not it would otherwise be damaged or destroyed by the elements or by vandalism.

Extracting a specimen embedded in a rock may prove to be a long and difficult process. Indeed, on occasion even the most beautiful specimen will have to be left behind since its collection would create too many problems, and even become hazardous. Before you attempt to collect a specimen make sure that it is feasible to remove it, and that the use of a hammer will not destroy either the specimen or the outcrop.

Your approach to the collection of a specimen will vary in relation to the rock type and the nature of the fossil. In hard, well-jointed rocks the best approach is to exploit the surrounding joints. Chisels can be inserted into the fractures to release the block concerned. Take care to use tempered steel tools and preferably chisels with a protected or strengthened head. Strike with the hammer working away from the body, and remember to wear goggles. Usually the fossils collected from such well-cemented rocks will not need any further preparation in the field. If it is impossible to exploit joints when extracting a specimen make sure that you chisel away from the specimen leaving a generous margin of sediment around it before trying to cut underneath. Be sure to leave sufficient rock beneath the specimen to protect against fracture.

Specimens collected from soft sediments may be delicately preserved. It is essential, therefore, that they are lifted in a block of sediment, so that their support is not removed until you get home. To do this it may be better to use a sharp knife rather than a hammer and chisel. Simply carve away the sediment. Remove the material at the side and back before you attempt to cut away the underlying sediment. When free, place the specimen in a box and support it with soft tissue paper.

When collecting from soft sands or sandy clays it may be better to sieve rather than scrape or dig into the material. In this way even the smallest specimens may be recovered and you will thus obtain a more balanced representation of the fossils preserved at that particular locality. Wet sieving is a technique frequently used for the collection of small mammalian fossils.

After collecting your samples number them immediately, and record the number and location in your notebook. This will prove important when you return heavily laden to home or laboratory. Make sure also to wrap your material carefully in tissue and newspaper. Never throw them all unnumbered into a box. Tissue paper, newspaper and strong tape are essential items for the palaeontologist in the field. Other items include a variety of small brushes for the removal of loose sediments and some polyvinyl-acetate (pva) glue for application to weakened specimens. Large specimens may need to be surrounded by a papier-mâché/plaster of Paris jacket before removal.

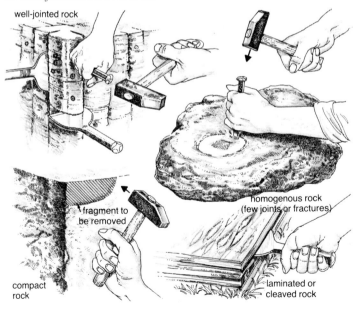

The technique to be employed in the extraction of a fossil will vary according to the characteristics of the rock. Do not hammer indiscriminately.

Back home

There are many techniques you can use to clean, prepare, preserve and enhance your fossil material. A good wash with water and a brush will often clean the more robust samples, but others should be treated with great care. Just brush these and pick away any loose sediment with a strong pin. A coat of thin polyvinyl-acetate glue will protect the exposed surface.

Some laboratories are well equipped with air abrasive tools, air dents and dentists' drills. With these it is possible to take away surplus sediment easily, but once again care is needed, especially when working close to the specimen.

Skilled preparateurs use a combination of these tools and diluted acids. The latter can be extremely dangerous in inexperienced hands so care is always needed when using these chemicals at home. Limestones respond well to treatment with dilute hydrochloric or acetic acids (2–10 per cent). Silicified fossils and bones can also be prepared from their sediments by using these chemicals. The specimens should be dry and the exposed parts of the fossils coated with a pva glue. Do not leave the specimen immersed in acid for too long – just four to six hours at a time. When you remove it, wash it by leaving it in a bath of clean water for an equal period of time and then dry it completely. The exposed areas of the fossil should again be coated before the process is continued. Many of the beautiful fossils on display in our great museums have been prepared this way.

Stages in the mechanical and chemical preparation of a body fossil. 1 – remove loose rock; 2 – immerse in acid; 3 – wash in water; 4 – paint with pva glue.

Curation and after care

When the preparation of your fossils is complete, it is worth taking the time to curate them properly. Make sure the number you placed on the specimen is transferred, together with the location details, to your collection book and specimen cards. A card should exist for each fossil. Apart from the location details, it should be marked with the drawer and tray numbers in which the specimen is kept. Other information, such as the name of the specimen, the family to which it belongs and the stratigraphic age should also be recorded. Establish a cross-index system between record book, cards and fossil and your collection will be a credit to you and science. Over the years you will be able to take a pride in it and compare your material with that of the major museums. On occasion, other palaeontologists will perhaps wish to visit you and use your material; specimens lacking all the details noted above will be of little or limited value. Update your information at regular intervals: names may change and outcrops disappear. If they do, record the details. Check also that the specimens are kept dry and that pyritized material does not decay. This is difficult to stop but soaking in a bacteriocidal disinfectant may arrest the process. Remember to dry the specimen concerned and coat it again with pva glue.

Naming and grouping your specimens

Unlike the vast majority of living plants and animals, fossils do not have popular names. Their names are derived from Latin or Greek and may be regarded as scientific names. Popular names such as 'dinosaur', 'clam' or 'ammonite' define groups, but not individual organisms. Scientific names are precise and are used to describe the characteristics of the fossil and define the group to which it belongs. Every animal or plant, however large or small, belongs to a **species** (in other words similar organisms that can interbreed to produce fertile offspring). Several species may belong to one **genus**. These will be similar in overall shape or character and are closely related. The scientific name of the organism consists of two parts – a **generic** name and a **trivial** or **specific** name. Intelligent man is therefore named *Homo sapiens* in accordance with this procedure. Genera are grouped into **families**, families into **orders**, orders into **classes** and classes into **phyla**. Each level of classification means a broadening of the characteristics that associate the constituent groups or **taxa**.

When you label your specimens it is essential that you give them the correct scientific name. For most specimens this will already exist, and can be found in books such as the *Treatise on Invertebrate Palaeontology*. Should you have difficulty in finding the right name then seek the help of your local museum or college. It may be that you have found something unique. If so it is necessary that you follow the correct procedure and describe and illustrate your specimen in a recognized scientific journal.

Identification keys

The world is constantly populated by potential fossils. After death, however, only a small percentage of living organisms will be preserved. Soft-bodied creatures will mostly vanish without trace, and even animals with hard parts will usually be destroyed or fragmented. The material preserved will therefore provide an unbalanced representation of life as we know it today. The same is true of the fossil record. It is incomplete, and many families have undoubtedly vanished without trace.

The recognition of organisms from fossil remains can be helped considerably by using a system of keys such as those on the following pages. Remember, however, that your specimen may consist only of a fragment of the complete animal, and so you must try and visualize the creature in its entirety. The first part of the key enables identification of a particular phylum or class to be made. Then, by turning to the relevant group in part 2, you can identify your fossil further by using a second key. Many types may be finally checked against the colour plates.

Part 1   Key to major fossil invertebrate groups

1	a Solitary	pass to 2
	b Colonial	pass to 3
2	a Non-chambered	pass to 4
	b Chambered (marked by septa)	pass to 5
3	a With pores	pass to 18
	b Without pores	pass to 19
4	a Coiled	pass to gastropods (page 45)
	b Non-coiled	pass to 6
5	a Coiled	pass to 7
	b Non-coiled	pass to 8
6	a Shell composed of single skeletal component	pass to 10
	b Shell composed of more than one skeletal component	pass to 13
7	a Chamber partitions straight or slightly curved	pass to 8
	b Chamber partitions folded	see ammonoids (page 108)
8	a Chamber partitions with small tube-like structure (the siphuncle)	pass to 9
	b Small to microscopic organisms; siphuncle absent	pass to foraminiferans (page 40)
9	a Chamber partitions (septa) straight or slightly curved	see nautiloids (page 106)
	b Chamber partitions slightly curved; major component of skeleton a bullet-shaped guard	see belemnites (page 118)

	c Chamber partitions folded	see ammonoids (page 108)
10	a Fossil characterized by radial symmetry	pass to 11
	b Fossil not characterized by radial symmetry	pass to 12
11	a With vertical radiating partitions	pass to corals (page 41)
	b Solid with central tube; radiate patterning	see crinoids (page 128)
12	a Fossil large with large aperture; pores present; single wall	pass to sponges (page 40)
	b Fossil large with large aperture; pores present; double wall	pass to archaeocyathids (page 41)
	c Fossil large with large aperture	pass to gastropods (page 45)
	d Fossil microscopic to small	pass to foraminiferans (page 40)
13	a Bilateral symmetry	pass to 14
	b Radial symmetry	pass to 16
14	a Shell composed of two valves	pass to 15
	b Segmented skeleton	pass to arthropods (page 48)
15	a Two valves, usually mirror images of each other and usually of equal size	pass to bivalves (page 46)
	b Valves usually of different size, not mirror images, equilateral symmetry	pass to brachiopods (page 43)
16	a Radial symmetry follows five-fold plan	pass to 17
	b Radial symmetry without five-fold plan	pass to corals (page 41)
17	a Skeleton with arms, and with or without stem	see crinoids (page 128) or blastoids (page 134)
	b Skeleton rounded or plate-like; arms incorporated into test	see echinoids (page 130)
	c Flattened; five radiating arms	see asterozoans (page 136)
18	With single porous wall	pass to sponges (page 40)
19	a Laminate structure; box-like units	pass to stromatoporoids (page 42)
	b Tube or box-like units	pass to corals (page 41)
20	a Large with vertical radial partitions	pass to corals (page 41)
	b Small to microscopic; no radiating vertical partitions	pass to 21
21	a Possessing rod-like branches	pass to graptolites (page 49)
	b Moss-like with many small apertures	pass to bryozoans (page 42)

Part 2　Key to fossil organisms

FORAMINIFERANS (Phylum Protozoa)
The majority of these single-celled creatures live in the sea. They frequently possess a skeleton or test that may consist of an organic, horny substance called chitin, of calcium carbonate or of sand grains cemented together around the animal. The test varies in size from 0.05 mm to 10 cm ($\frac{1}{50}$ in to 4 in). Specific families have a characteristic test wall structure. Shape and size are good indicators of mode of life, whereas the number and arrangement of the chambers help in the classification of genera and species. You will need a powerful microscope to find and study the smaller foraminiferans. Larger forms, however, exist throughout the stratigraphic record. Some of these are important rock formers and their abundance makes them easier to find (see page 52).

SPONGES (Phylum Porifera)
Sponges are many-celled organisms that rank just above the protozoans in order of classification. The majority are marine animals that range in size from less than 1 cm ($\frac{1}{2}$ in) to one metre in diameter. Although they have many cells the sponges have no recognized tissue layers. The soft body is often supported by thin, rod-like elements termed **spicules**. These may be separate or fused, and composed of either calcium carbonate, silica or a horny substance called **spongin**. Individual, unfused spicules are difficult to find or recognize, but skeletons composed of welded or cemented elements are common to a number of Mesozoic and Cainozoic horizons. The mineral composition of a sponge skeleton is an indication of both its classification and its life style. It is possible with the descriptions given below to identify the group to which your specimen belongs. A number of genera are illustrated, and these represent the most common fossil sponges.

Key to major groups of fossil sponges

1	a	Skeleton calcareous	pass to 2
	b	Skeleton composed of siliceous spicules	pass to 3
2	a	Skeleton with thin walls; cylindrical in shape; small osculum	example: *Peronidella* (page 56)
	b	Skeleton with thick walls; solitary or colonial; cup or vase shaped with distinct central cavity	example: *Raphidonema* (page 56)
3	a	Skeleton cup or vase shaped, rarely branched; spicule arrangement box-like, spicules six-rayed	example: *Ventriculites* (page 54)
	b	Skeleton massive, densely constructed with small or	example: *Siphonia* (page 54) or *Entobia* (page 54)

reduced openings; spicules
large, knobbly; sponges
sometimes leave burrow or
boring as a trace

ARCHAEOCYATHIDS (Phylum Archaeocyathida)

A very distinctive group of invertebrates known only from the Lower
and Middle Cambrian. They are similar to sponges in that they possess
pores, but their skeletons are always cup shaped and usually have a cup
within a cup. The two are separated by horizontal and vertical partitions.
All walls and subsidiary skeletal elements are pierced by pores. A root
structure may be found in well-preserved specimens.

CORALS, JELLYFISH AND THEIR RELATIVES (Phylum Cnidaria)

These are the simplest of animals which have cells organized into tissues.
Their tissues are organized on a radial plan, and this is often reflected in
the hard parts of various families.

The most common fossil representatives of the Cnidaria are the stony
corals. These have a cylindrical body with a central mouth surrounded
by a ring of tentacles. The animal is described as a **polyp**. It contrasts
with the discoidal body, with downward-facing mouth, of the
free-swimming jellyfish **medusa**. Sea-anemones are typical examples of
fixed polyps. The stony coral polyps sit on top of a calcareous skeleton.
They belong to the class Anthozoa.

The calcareous skeleton is produced and deposited by the soft tissues at
the base of the polyp. Originally the skeleton is aragonitic but most fossil
corals are composed of recrystallized calcite. The growth of the animal is
reflected in the development of the skeleton. The latter is essentially
horn-shaped with an outer wall and numerous radially arranged septa. It
is termed the corallum. The number of septa, the way they develop, and
the form and variety of other skeletal elements are important in the
classification of the stony corals. Single (solitary) and colonial corals can
be recognized within the fossil record. The following key enables you to
identify your coral specimens to order level. The 'stony' corals' belong to
three orders. Of these the tabulate and rugose corals are known only
from the Palaeozoic; the scleractinians or hexacorals are known only
from the Mesozoic and Cainozoic. Related groups such as the
Stromatoporoidea (*see* below) and the Hydrozoa (*see* page 68) are also
important during the Palaeozoic and Cainozoic respectively.

Key to major groups of fossil corals

1	a	Septa poorly developed or absent	pass to 2
	b	Septa well developed	pass to 3
2		Horizontal partitions (tabulae) the dominant structural element	see tabulate corals (page 58)
3	a	Solitary or colonial corals with well-developed tabulae	see rugose corals (page 60)

and septa; confined to
Palaeozoic
b Corals with developed septa   see scleractinian corals
and tabulae; confined to   (page 64)
Mesozoic and Cainozoic

STROMATOPOROIDS (Class Stromatoporoidea)
Found mostly as layered calcareous masses, the stromatoporoids were
important reef-builders during the Palaeozoic. The skeleton may take the
form of rounded masses or thin sheets, and is rarely branched or
cylindrical. It is often difficult to observe the finer structure of the
skeleton but stromatoporoids do resemble certain tabulates. The surface
of the colony is frequently mammelate, with a central opening marking
the position of the animal. Numerous grooves with a star-like
distribution radiate outwards from the opening. It is likely that the
extinct stromatoporoids are related to the living reef-building hydrozoans such as *Millepora*. Like the 'stony corals' these have a polyp-shaped
body. The stromatoporoids range from the Cambrian to the Cretaceous.

BRYOZOANS (Phylum Polyzoa)
Although small and rather delicate the bryozoans are commonly found as
fossils. They are colonial animals, the majority of which exist or existed
in marine environments. The colonies consist of numerous small tubes or
boxes and superficially they resemble the smaller tabulate corals. In fact,
the soft parts are characterized by the presence of a mouth, gut and anus
and the animals have a closer affinity with the brachiopods than with the
more primitive corals. The skeleton may consist of several sizes of tubes:
the larger are termed **autozooecia** (**autopores**), the smaller **meso-
zooecia** (**mesopores**) and **ancanthostyles**. Horizontal plate-like
structures called diaphragms subdivide the tubes. They may be
compared with the tabulae of corals, but septa and dissepiments (small
calcite plates) are absent. In the more advanced genera a frontal wall
partially seals the upper opening of the tube or box. The first bryozoans
appeared in the Ordovician and over 3,500 species exist at the present
time. The phylum is subdivided into three classes of which the
Stenolaemata and Gymnolaemata are most important.

The Stenolaemata includes the calcareous bryozoans. They are
organized into four orders, the classification of which is based on the
structure and organization of the colony. The Gymnolaemata includes
both calcareous and horny bryozoans.

Key to major groups of fossil calcareous bryozoans
1 a Autozooecia dominant feature   pass to 2
  b Autozooecia separated by   pass to 3
    intervening mesozooecia
    and/or ancanthostyles
2 a Autozooecia rounded without   example: *Berenicea*,
    frontal wall   *Theonoa*, *Meandropora*
                                      (page 70)

	b Autozooecia box-like with complex frontal wall over aperture	cheilostome bryozoans (not illustrated)
3	a Colonies delicate	pass to 4
	b Colonies massive; stony	trepostome bryozoans (not illustrated)
4	a Zooecia connected by mural pores; tubes elongate, tubular	cryptostome bryozoans (not illustrated)
	b No mural pores; zooecia short, cylindrical	example: *Fenestrella*, *Archimedes* (page 70)

BRACHIOPODS (Phylum Brachiopoda)

Brachiopods are sea floor dwellers. Their soft parts are protected inside a two-valved chitinophosphatic or calcareous shell. Superficially, the shell resembles that of a bivalve, but the individual valves are symmetrical about a medial plane. They are also of unequal size. In contrast, the shell of a bivalve consists of two equal-sized valves that are not symmetrical about a median plane. Two major groups can be recognized on the basis of shell composition and structure.

INARTICULATES (Class Inarticulata)

Brachiopods in which the shell is usually chitinophosphatic, rarely calcareous. The valves are not hinged by teeth and sockets, and internal support structures are absent. The chitinophosphatic nature of the shell can often be determined by colour and texture. Chitinophosphatic valves are usually brown-black or black, lustrous and with a layered horny appearance.

Key to major groups of fossil inarticulate brachiopods

1	a Shell chitinophosphatic (black/ brown in colour with strong lustre; wall of shell very thin)	pass to 2
	b Shell calcareous (mineralized; generally not as above)	
	(i) small with indistinct hinge line	pass to 2
	(ii) medium to large with distinct hinge line	pass to articulates (page 44)
2	a Valves elongate, subequal	pass to 3
	b Valves rounded or subrounded	pass to 3
	c valves circular or subcircular	pass to 3
3	a Pedicle emerges between valves	example: *Lingula* (page 72)
	b Pedicle opening through pedicle valve only	pass to 4
4	a Shell phosphatic, round or oval	example: *Orbiculoidea* (page 72)
	b Shell phosphatic or	example: *Crania* (page 72)

calcareous, circular or
subcircular

c Shell calcareous, biconvex,   example: *Discina* (not
circular or oval   illustrated)

5 a Shell calcareous with   example: *Kutorgina* (not
prominent beak on pedicle   illustrated)
valve, hinge line well
developed for inarticulate

ARTICULATES (Class Articulata)
Brachiopods of moderate to large size with calcareous shells. Hinge line
well developed, with teeth and sockets for articulation. Internal support
structures may be extended to form spiral or loop-like features. External
characters and ornament will help in a rudimentary identification.

Key to major groups of fossil articulate brachiopods

1 a Hinge line straight, equal to   pass to 2
width of shell (strophic)

   b Hinge line curved or straight,   pass to 6
narrower than width of shell
(non-strophic)

2 a Shell biconvex   pass to 3

   b Shell plano-convex   pass to 3

   c Shell concavo-convex   pass to 3

3 a Finely ribbed   pass to 4

   b Strongly ribbed   pass to 5

4   Pedicle aperture open   see orthids (page 74)

5 a Pedicle aperture closed   see strophomenids (page 82)

   b Pedicle aperture open   see spiriferids (page 84)

6 a Shell biconvex, strongly ribbed   pass to 7

   b Shell biconvex, smooth or   pass to 8
weakly ribbed

7 a Shell corrugated, folded with   see rhynchonellids (page 76)
strong beak

   b Shell without fold   pass to 8b

8 a Shell subcircular to elongate   see terebratulids (page 80)
tear-drop shape; prominent
rounded pedicle opening

   b Shell strongly biconvex, beak   see pentamerids (page 78)
well developed on one or both
valves

The characters used in our key are restricted to the external surface of
the shell. Detailed descriptions of each order should also contain mention
of internal characters.

MOLLUSCS (Phylum Mollusca)
Living molluscs include snails, slugs, clams (bivalves), squids and
chitons. They represent a very diverse range of animals, and when

considered together with the extinct groups constitute the most important group of fossil animals.

The molluscs occupy many ecological niches and fulfil many ways of life. The clams (bivalves) may move freely or burrow and bore into the substrate. Snails crawl and scavenge, whilst squids and associated forms, swim and hunt in the open seas. It is likely that all molluscs evolved from a common ancestor in Precambrian times. The living chitons may be closest to the possible ancestor. They have a segmented shell and a rather simple body structure.

Fossil molluscs are invariably recognized by their shells. These vary in structure and tell us much about the form and habits of their original occupant. Molluscan shells are composed of calcium carbonate, either in the form of calcite or aragonite, which in many genera is covered by a thick, dark coloured, organic layer called the periostracum. The latter is rarely found fossilized. Various features exist that enable you to make a rapid determination to class level. From there, specific characters such as ornament, the organization of teeth and sockets, the type of coiling and the form of the suture between the inner and outer shell wall are critical to the identification of families and of individual genera.

Key to major groups of fossil molluscs

1	a	Shell segmented	see chitons (page 104)
	b	Shell unsegmented	pass to 2
2	a	Shell coiled	pass to 3
	b	Shell uncoiled	pass to 5
3	a	Shell chambered	pass to 4
	b	Shell non-chambered	pass to gastropods (page 45)
4	a	Chamber partitions folded	see ammonoids (page 108)
	b	Chamber partitions straight or slightly curved	see nautiloids (page 106)
5	a	Shell composed of fused single unit	see scaphopods (page 104)
	b	Shell composed of two valves	pass to bivalves (page 46)

GASTROPODS (Class Gastropoda)
One of the main divisions of the Phylum Mollusca, the gastropods (snails and slugs) are characterized by a univalved shell. It is calcareous and, in the majority of gastropods, coiled. In the simplest forms this occurs around a single plane and is termed **planispiral**. The shell can be divided into two halves, each the mirror image of the other. This is not true for the majority of shells coiled helically, in which the coiling moves away from the plane to resemble a wire wound around an upright cone. The height of the cone may vary from low or depressed to tall or elongate. Each complete turn of the gastropod shell is called a **whorl**. The main whorl contains the body of the animal. On the outside the whorls are separated by sutures and ornamented by growth lines, ribs, nodes and spines.

Key to major groups of fossil gastropods

1	a Shell present	pass to 2
	b Shell absent or reduced	opisthobranchiates (not illustrated) and pulmonates (page 92)
2	a Shell coiled	pass to 3
	b Shell uncoiled	amphigastropods (not illustrated)
3	Shell coiled planispirally	amphigastropods (not illustrated) and archaeo-gastropods (page 86)
4	a Without siphonal canal	pass to 5
	b Mostly with siphonal canal	pass to 7
5	a Indentation on outer edge of aperture (slit-band) frequently present	see archaeogastropods (page 86)
	b Aperture entire, often rounded; no slit-band	pass to 6
6	a Shell thin; delicate coiled or cap-shaped	opisthobranchiates (not illustrated) and pulmonates (page 92)
	b Shell robust or moderately robust; mostly smooth but often with modified inner lip or aperture	see mesogastropods (page 88)
7	Siphonal canal prominent, often elongate or extended	see neogastropods (page 90)

BIVALVES OR CLAMS (Class Bivalvia)

The bivalves have a shell composed of two calcareous valves. These enclose the soft parts of the body. Most clams are marine but freshwater genera are quite common in some lakes and rivers. Boring, burrowing, cemented and free-living clams are abundant in shelf environments at the present day. During the Palaeozoic, brachiopods were the most common elements of shallow marine faunas, but from the Upper Triassic onwards bivalves are dominant.

The valves of the majority of bivalve shells are marked by concentric growth lines and many have pronounced ornamentation. This may take the form of ribs or spines. Where the valves are of equal size they are termed equivalve, whilst those of different proportions are termed inequivalve. It is possible to identify a right and left valve for each shell on the basis of specific characters. The hinge line is usually dorsal in position, the shell opening or commissure ventral. A number of teeth and socket arrangements occur along the hinge of different bivalves. These, together with the muscle scars, are useful clues to function and habitat and may be important in the classification of the class. Several classifications of the Bivalvia exist, and we have presented an identification key based on shell characters.

Key to major groups of fossil bivalves

1	a	Shell with teeth	pass to 2
	b	Shell toothless or teeth very weak	see anomalodesmates (page 102) and pteriomorphs (page 94)
2	a	Teeth numerous, subequal in size, parallel to each other; oblique or vertical to hinge line	pass to 3
	b	Teeth few in number and of different shapes	pass to 4
	c	Teeth vertical or oblique; lateral teeth parallel to outline of hinge	see heterodonts (page 98)
3	a	Shell small, thin and of conventional shape	see palaeotaxodonts (page 94)
	b	Shell variable in shape; equivalve or inequivalve, thick with variable-sized muscle scars	see pteriomorphs (oysters and mussels) (page 94)
4	a	Hinge or cardinal teeth only partially differentiated from lateral teeth; number of teeth may be reduced to two or three	see palaeoheterodonts (page 98)
	b	Hinge teeth and lateral teeth clearly differentiated; number of teeth may be reduced or shells may be modified to toothless condition	see heterodonts (page 98) and pachydonts (page 102)

CEPHALOPODS (Class Cephalopoda)

These are the most highly developed of all molluscs. Living and fossil species are characterized by the presence of a prominent head and well-developed sensory systems. The brain is large by invertebrate standards and the eyes have structural similarities with those of the vertebrates. Squids, cuttlefish and octopuses are living examples of this class. Another, rarer representative is the pearly *Nautilus* which is confined territorially to the Indo-Pacific area. It has a coiled, external shell similar to those of a myriad fossil species. The shell is divided by walls or **septa** into chambers or **camerae**, that are filled with gas. A tube, the **siphuncle**, connects the chambers. The siphuncle, which passes through each septum, controls the fluid/gas levels in the chambers and is therefore closely associated with the buoyancy of the animal. The line of contact between the septum and the shell wall is termed the **suture**. The latter may be simple as in *Nautilus* or complexly folded as in the ammonites. Suture lines are important in the classification of several of the cephalopod groups.

Key to major groups of fossil cephalopods

1	a	Shell divided into three components	see belemnites (page 118)

		b	Shell single structure, coiled or uncoiled	pass to 2
2	a		Suture line straight, or gently curved	see nautiloids (page 106)
		b	Suture line folded	pass to 3
3	a		Suture line folded into broad, open folds	see goniatites (page 108)
		b	Suture line marked by folds subdivided into smaller secondary folds	see ceratites (page 110)
		c	Suture line complexly folded	see ammonites (page 112)

The broad classification given above is useful in terms of a general field identification. In order to classify specimens to generic and specific level it will be necessary for you to refer to the *Treatise on Invertebrate Palaeontology*. The genera illustrated in this book are, however, representative of the major families of cephalopods and the detailed descriptions serve as examples for your own descriptive notes.

ARTHROPODS (Phylum Arthropoda)

Of all the invertebrates the arthropods are perhaps the most spectacular. They possess a distinctive segmented outer skeleton (**exoskeleton**) and jointed limbs. Members of the group have occupied many niches and thrive in most environments. They include the insects, spiders, crabs, lobsters and microscopic ostracodes. Their tough exoskeleton protects them from many predators and enables them to survive the harshest of conditions. It also increases their potential for fossilization, and the records of both insects and crustaceans (crabs and lobsters) can be traced back to the Lower Palaeozoic. The dominant group of arthropods during the Palaeozoic are now extinct. They were called trilobites, and were in fact the earliest known arthropods. They are characterized by a three-part exoskeleton and jointed limbs. The skeleton is subdivided across and along its length into three distinct areas. The main division is into head (cephalon), body (thorax) and tail (pygidium). The head is usually semicircular in shape with a central raised area called the glabella. The eyes are situated on the flat or slightly domed areas on either side of the glabella. They are made up of numerous box-like or rounded lenses and usually take the form of prominent crescentic ridges. Hairline sutures are frequently observed to pass from the front margin of the cephalon, around the eye to cut either the lateral or posterior margins. They are called the **facial sutures**. The thorax is divided into many segments with a prominent hump, the axis, occurring along its length. In many trilobites the tail is also semicircular. The segments may be fused together to form a single plate. A central axis is also present in many specimens. The existing classification of the trilobites is based on many characters. These may be difficult to recognize in the field and therefore an older classification based on the type of facial suture is used as the basis of our key.

Key to major groups of trilobites

1	a	Facial sutures present	pass to 2
	b	Facial sutures absent	pass to 3
2	a	Facial suture marginal; eyes large, crescentic and sometimes joined to glabella; tail small, spiny	example: *Olenellus* (page 122)
	b	Facial suture cuts side of head shield	example: *Olenus* (page 124)
	c	Facial suture cuts posterior margin of head shield	example: *Ogygiocaris* (page 124)
	d	Facial suture marginal; no eyes; small rounded tail	example: *Trinucleus* (page 122) example: *Peronopsis* (page 122)
3		Small trilobites with two or three body segments; head and tail roughly equal in size	agnostids and eodiscids

OSTRACODES (Class Ostracoda)

Known since the Ordovician the ostracodes are an important group of microscopic organisms. They possess jointed limbs typical of other arthropods, but have a distinctive two-valved outer covering. The majority of shells fall within the range of 0.5–5 mm, but a few species reach 2 cm ($\frac{3}{4}$ in) in length. The valves are joined along a straight hinge line, with hinge teeth present on one valve and sockets on the other. The external ornament and the overall shape of the shell is important to the identification of individual genera and species. Ostracodes occur today in marine and freshwater environments.

GRAPTOLITES (Class Graptolithina)

The graptolites are colonial marine animals that first appeared in the Cambrian. They are related to the chordates, their delicate 'fabric' skeletons resembling those of the living pterobranch *Rhabdopleura*. The fabric sheets are probably organic but most skeletons are found as carbonized or pyritized remains. Often you will only recognize the skeleton as a toothed, blackened branch, but sometimes well-preserved specimens will exhibit a number of diagnostic features. The most obvious will be the branches or **stipes**. The toothed appearance of the stipe is due to the presence of cup-like **thecae**, which may vary in size and number. Each graptolite family is characterized by a given number of stipes. The shape and arrangement of the thecae are also important in the recognition of specific genera. It is possible to divide the graptolites into two major groups: the orders Dendroidea and Graptoloidea. The former have two types of thecae: small **bithecae** and large **autothecae**, and numerous stipes. In contrast, the graptoloids possess only the larger autothecae and the number of stipes rarely exceeds eight.

Throughout the Ordovician and Silurian there is a progressive reduction in the number of stipes in successive graptolite families. Graptolites with 8, 4, 2 and 1 stipe may be recognized in various sediments. Many species are short lived but widespread. They make excellent zone fossils.

Key to major groups of graptolites

1 a Stipes numerous; colony shrubby in appearance: show two types of cup under high magnification — see dendroid graptolites (page 138)

  b Stipe number limited to 8 or less — pass to 2

2 a Stipe number 5–8 — *Dichograptus* (not illustrated)

  b 4 stipes — example: *Phyllograptus* and *Tetragraptus* (page 138)

  c 2 stipes — example: *Didymograptus* (page 138)

  d 1 stipe — pass to 3

3 a Thecae on both sides of stipe — example: *Diplograptus* and *Dicranograptus* (page 138)

  b Thecae on one side of stipe — example: *Monograptus* and *Rastrites* (page 140)

ECHINODERMS (Phylum Echinodermata)

Most of us are familiar with the starfish and brittlestars that live in coastal waters. They have an internal skeleton composed of calcite **plates**. The animal has a central disc-shaped body and five arms. (Five is a number commonly associated with the echinoderms, as many groups exhibit **pentameral symmetry**, or have a five-rayed configuration.) In the starfish and brittlestars the plates are not fused together and as a consequence they fall apart after death. Other echinoderms, such as the sea-urchins, have robust, rounded or discoidal tests with fused or connected plates. They are common fossils in many carbonate sediment sequences.

The echinoderms are varied in form and function, and the overall structure of the skeleton will reflect the mode of life. The starfish and sea-urchins are burrowing and free-moving scavengers. In contrast, the sea-lilies are mostly fixed sea-floor dwellers: a stem and rooting structure anchoring them to the substrate. A common feature among echinoderms is the presence of an internal complex of tubes and bladders. This is termed the water vascular system and may be recognized in living animals as the tubular extensions which aid feeding, movement and respiration. The tubes are called tube-feet, and their presence in fossil species is recognized by a sequence of **pores** in specific plates. Each group of echinoderms has its own characteristic features; these are also important in the identification and classification of the individual specimen.

Key to major groups of fossil echinoderms

1 a Skeleton with stem or stalk-like process — pass to 2

b	Skeleton without stem	pass to 5
2 a	Skeleton symmetrical	pass to 3
b	Skeleton without symmetry	pass to 4
3 a	Pentameral symmetry pronounced; small, bud-like cup or calyx consisting of 13 plates	see blastoids (page 134)
b	Pentameral symmetry; calyx with regularly arranged rings of plates	see crinoids (page 128)
4 a	Plates of calyx irregularly arranged; several pore types	see cystoids (page 134)
b	Plates regularly or irregularly arranged; pores limited in number or absent	see carpoids (page 136)
5 a	Skeleton with disc-shaped body and five distinct arms	see asterozoans and ophiuroids (page 136)
b	Skeleton or test rounded; entire, with five pore-bearing areas each two plates wide	see echinoids (page 130)
c	Skeleton cup-shaped with plates arranged in rings around cup	see crinoids (page 128)

PLANTS AND VERTEBRATE FOSSILS

The majority of plant remains are found as casts or moulds, or as carbonaceous films on silicified lumps of wood. The casts and moulds of tree trunks, roots and branches are common in sandstones associated with coal deposits. Impressions of leaves and fronds are frequent in the actual coal seams, along with the carbonaceous remains of the more woody parts. Leaf beds are common in the Mesozoic and Cainozoic, and experts can identify the species on leaf shape and venation. Fossil forests are also more common in Mesozoic and younger strata, and a detailed analysis of the woody tissues will provide a key to identification. A number of fossil plant groups are described and illustrated on pages 148 to 155. For a more detailed understanding of fossil plants, however, you should consult specialist text books and visit a museum.

The same is also true for vertebrate remains, as one animal may be known by hundreds of bones. Sharks can be readily identified by their sharp pointed teeth and individual mammals on the shape and size of individual canine or cheek teeth. Fish bones are often shiny brown or black in colour, although no hard and fast rule exists for easy identification. Books such as *Palaeozoic, Mesozoic* and *Cainozoic Fossils* published by the British Museum (Natural History) will help with the identification of the more common discoveries. It is recommended, however, that large, intriguing and important specimens are reported to your local museum. There the curators of the vertebrate and plant collections will be able to help in the determination of your material.

Some fossil vertebrates are shown on pages 142 to 147.

Foraminiferans

Microscopic to small unicellular animals which secrete a calcareous test (shell). This may consist of one or several chambers, and may be coiled or uncoiled. Range Cambrian to Recent.

Fusulinids   The most characteristic feature of the majority of fusulinids is the spindle-shaped test. Fusulinids are multichambered, the chambers being arranged around an axis of coiling. They are very common in rocks of Upper Carboniferous and Permian age. Fusulinids are commonplace in the mid-continental region of the USA and in China.

Nummulites   These larger foraminiferans are essentially disc-shaped and multichambered. The test is composed of calcium carbonate and perforated with small holes. The outer surface may be marked with a curved or spiral ornamentation and sometimes with small nodules. Nummulites are first found in Palaeocene rocks and are most common in areas bordering the Mediterranean and Caribbean seas; they became extinct in the Oligocene. Nummulites lived in warm, tropical seas.

Orbitoids   The orbitoids are grouped with the larger foraminiferans and have a disc-shaped test composed of calcium carbonate. Numerous chambers are characteristic, their general arrangement being in rings around a limited number of central chambers. The orbitoids range from the Cretaceous to the Miocene, and lived in warm, shallow seas.

Globigerinids   The globigerinids are a group of planktonic foramini-ferans. They first appeared in the Middle Jurassic, and exist today in most seas. The name globigerinid is often used as a 'sack' name to describe a number of related genera and species including *Globotruncana*, *Globorotalia* and *Globigerina*. The planktonic habits of these animals ensure a widespread distribution, and short-lived species are used for the correlation of rocks over large areas.

Textulariids   Unlike the other groups of the Foraminifera noted above, the textulariids have agglutinated tests. This means that foreign particles (from the sea floor or animal remains) are stuck together to form a rigid skeleton around the animal. They live today on the sea floor in shallow coastal waters or on the abyssal plains beyond the continental shelf. Agglutinated foraminiferans first appeared in the Cambrian.

Quinqueloculinids   These are essentially microscopic foraminiferans. They belong to the Order Miliolida, the calcareous tests of which have a shiny, porcellaneous appearance. The quinqueloculinids have a rather angular appearance with only five chambers visible on the outside surface. They range from the Jurassic to the present day and are good indicators of back reef, lagoonal environments.

Fusulinids
(×1·6)

Nummulites
(×1)

Orbitoids
(×1·6)

Globigerinids
(×19)

Textulariids
(×14)

Quinqueloculinids
(×14)

Sponges

Multicellular animals with calcareous or siliceous skeletons comprised of single or many-rayed **spicules**; these may be fused to form a rigid framework. Solitary and colonial genera are recorded. Range Precambrian to Recent.

Sponges with siliceous skeletons

Siphonia   This genus is a member of the demosponges. The main body is globular in shape with a pronounced, but small to medium-sized, opening on the upper surface. *Siphonia* was a stalked sponge. The stalk is long and slender; the whole skeleton having a tulip-like appearance. The genus is common in the Upper Cretaceous of Europe. Its range is Middle Cretaceous to Tertiary and it is confined to Europe.

Doryderma   Like *Siphonia*, *Doryderma* is a demosponge, the large spicules (desmas) giving the surface a rather lumpy appearance. The skeleton has a branched, plant-like shape, the numerous cylindrical branches arising from a slender stalk. Each branch has its own opening or **osculum**. *Doryderma* ranges from the Carboniferous to the end of the Cretaceous, but it is confined to Europe throughout its history.

Ventriculites   This is a member of the group Hyalospongea or 'glasssponges'. Its skeleton is composed of six-rayed spicules that fused to give a rigid framework. The skeleton is vase-shaped with a large osculum on the upper surface. Radiating 'root-like' structures fixed *Ventriculites* to the substrate. This genus is found only in the Middle and Upper Cretaceous rocks of Europe.

Seliscothon (Laosciadia)   A small to medium-sized siliceous sponge in which the skeleton is made up of rather smooth spicules and small-rayed desmas. The skeleton is vase or funnel-shaped. *Seliscothon* and its relatives are most abundant in sediments of Cretaceous age. The genus *Seliscothon* is restricted to Europe.

Entobia (Cliona)   *Entobia* is a burrowing demosponge, the soft parts of which are bound by a leathery cover and the spicules are loosely arranged. The meandering galleries formed by the sponge are abundant at certain horizons, however, particularly where silicified. *Entobia* is found worldwide and ranges from the Devonian to Recent.

Plocoscyphia   This attractive sponge is composed of six-rayed spicules. It is related to *Ventriculites* but has a complex skeleton containing many convoluted tubes, some of which branch and fuse to give a plant-like appearance. *Plocoscyphia* is known from Europe and has a Jurassic to Cretaceous range. Good material may be collected from the Chalk Marl (Cretaceous) of the Isle of Wight, England.

Siphonia

Doryderma

5 cm

Ventriculites

Seliscothon

Entobia

Plocoscyphia

Sponges with calcareous skeletons

Corynella   This small sponge is characterized by a skeleton made up of three-rayed, tuning-fork shaped spicules. The spicules form a rigid framework which is cylindrical in form, with a well-defined osculum. The outer surface may have a knobbly appearance but the majority of specimens are quite smooth. The species *C. guensted* is common in the Upper Jurassic of Germany, and *C. foraminosa* is abundant in the Lower Cretaceous of England. *Corynella* ranges from the Triassic to the Cretaceous, and is recorded from Europe and possibly the East Indies.

Raphidonema   Like *Corynella* this genus has a rigid skeleton composed of three-rayed spicules. Both belong to the family Pharetronida. *Rhaphidonema* is characterized by erect, vase-shaped or inverted cone-like skeletons. The outer surface is rough to the touch and usually lumpy or mammelate. Large pores are visible on the outer surface. *R. farringdonense* is named after the town of Farringdon in England where sponges are commonly found. This sponge is confined to Europe, and ranges between the Triassic and the Upper Cretaceous.

Peronidella   This small to medium-sized calcisponge is known from the Mesozoic sediments of Europe and the West Indies. It is a colonial sponge with numerous, cylindrical, finger-like columns. The main opening, or osculum, occurs at the tip of each column. Some species are more organ-like in appearance, the columns arising from a common base. Excellent examples are recorded from the Lower Cretaceous Farringdon Sponge Gravels of England.

Oculospongia   is a phraretone calcareous sponge. These terms relate to the linear arrangement of the skeletal elements. The resultant skeleton is rigid in character, with the erect cylinders having a rounded, often cushion-like appearance. The main opening is large and deep. *Oculospongia* is recorded from the Lower Cretaceous sediments of Europe.

Porosphaera   This is a small, rounded sponge in which the pores and oscula are distributed over all of the outer surface. The latter may have a spiny appearance under the microscope. Many specimens occur in the chalky sediments of the Cretaceous; the majority having a small, spherical shape. *Porosphaera* is known from the Cretaceous of Europe, but related genera are cosmopolitan in their distribution.

Corynella

Raphidonema

5 cm _____

Peronidella

Oculospongia

Porosphaera

Corals

These marine animals possess a soft, polypoid body similar to that of the sea-anemone. They deposit calcareous tube-like *corallites* that combine in colonial genera to form a **corallum**. Three major orders exist. Range Ordovician to Recent.

Tabulate corals

Colonial corals with relatively few skeletal elements. Septa poorly developed; horizontal tabulae well represented in many genera. No central rod. Genera confined to Palaeozoic rocks.

Favosites A colonial coral with slender corallites. These are five-sided, having short spinose septa and interconnecting pores. The tabulae are closely spaced and strongly developed. *Favosites* is first found in Middle Ordovician rocks. It is particularly abundant in Silurian strata.

Aulopora A colonial coral often found encrusting other organisms or inorganic surfaces. The corallites are short and slightly curved. They form a chain, with corallites cemented in 'head-to-tail' fashion. The trumpet-shaped corallites contain poorly developed septa. Ranges worldwide from the Silurian to the end of the Carboniferous.

Syringopora The long, cylindrical, and irregularly branched corallites of *Syringopora* form large colonies. Some species are characterized by the presence of 12 slender septa. Tabulae are numerous and well developed. *Syringopora* ranges from Silurian to Upper Carboniferous.

Heliolites This rather unique tabulate coral consists of large corallites, similar to those of *Favosites*, bounded by at least 12 small, tube-like structures. The corallites have small, spinose septa and well-developed tabulae. Similar horizontal partitions occur across the thin-walled tubes. The coral is cosmopolitan in distribution during the Middle Palaeozoic.

Thamnopora (Pachypora) This unusual tabulate coral forms massive, erect colonies. These are composed of short, closely packed corallites that branch frequently. Short, spinose septa and thin tabulae are present. *Thamnopora* is found worldwide from the early Silurian to the end of the Permian.

Halysites This tabulate coral is commonly known as the 'chain-coral'. The colonies consist of numerous elongate corallites cemented together along the entire length of their adjoining edges. The upper surface of the colony has a chain-like pattern. Individually the corallites are ovoid in outline with no recognizable septa. In longitudinal section, tabulae are common. They are slightly arched in character. *Halysites* is found almost worldwide but is essentially confined to the Silurian Period.

Favosites

Aulopora

Syringopora

Heliolites

Thamnopora

Halysites

5 cm

Rugose corals

Solitary or colonial 'stony' corals with well-developed septa, dissepiments and tabulae. Range Ordovician to Permian. Worldwide.

SOLITARY RUGOSE CORALS

Petraia   This coral has long thin septa extending to the centre of the corallite. Minor, small septa are prominent and they alternate with the major, longer septa. Cup or calyx deeply bowl-shaped. *Petraia* has few tabulae and no dissepiments. Upper Silurian of the Welsh borders in Great Britain, and elsewhere in Europe.

Zaphrentis   Small to medium-sized coral of Lower Carboniferous shallow marine faunas. The corallite is horn-shaped with elongate major septa; these meet just outside the centre of the cup, with a pronounced interseptal space present on the inside (curved) area of corallite. The minor septa may be reduced or absent. In some species the major septa are also shortened. *Zaphrentis* and related genera are recorded from the Devonian of Kentucky, USA, and the Lower Carboniferous limestones of Europe and Asia.

Caninia   A medium to large-sized coral which is sometimes cone-shaped, but many specimens are cylindrical. The external surface of the corallite is roughly textured, with distinctive growth lines providing the ornament. In adult specimens the septa are short and dilated. Dissepiments are present toward the outside of the corallite. The tabulae are numerous and flat. *Caninia* is recorded from Europe, North America, Asia and Australia.

Dibunophyllum   This, like *Caninia*, is another medium to large-sized solitary coral. It is distinguished, however, by the presence of a complex axial area. This occupies approximately one-third of the area of the cup and is web-like in character. *Dibunophyllum* has a large number of septa and numerous dissepiments. It is known from the Lower and Middle Carboniferous sediments of Europe, North America and North Africa.

Calceola   Often called the 'slipper coral', *Calceola* is one of the most distinctive solitary corals. The corallite is small, with a flattened under-face. It tapers to a rounded point and is slightly curved in this lower region. *Calceola* has a lid-like operculum. The cup or calyx is shallow, with the major septa of the upper and lateral edges being longer than those of the lower one. *Calceola* is known worldwide from the Lower and Middle Devonian. Excellent examples are recorded from Michigan, USA, and the Ardennes in Belgium.

Goniophyllum   A small solitary coral in which the corallite is square in cross-section. The septa are thickened, and dissepiments are common nearer the outer wall. *Goniophyllum* has a lid-like operculum consisting of four triangular plates. It is confined to the Silurian rocks of Europe.

Petraia

Zaphrentis

Dibunophyllum

Calceola

Goniophyllum

5 cm

Caninia

COLONIAL RUGOSE CORALS

Hexagonaria   This coral occurs in large to massive colonies. The individual corallites are tightly packed, with contact along all five or six sides. Small lateral outgrowths (carinae) occur on the strongly developed major septa. The septa terminate short of the axis and in weathered specimens this is often shown as a circular, central pit. *Hexagonaria* is abundant in shallow-water Devonian limestones. It is known worldwide and may have persisted into the Carboniferous in Asia.

Lithostrotion   This is one of the best-known Carboniferous corals. Numerous species occur worldwide, with the large to massive colonies abundant in shallow-water Lower Carboniferous limestones. The forms of the colonies vary according to the shape and arrangement of the corallites. In *L. junceum* they are circular in outline, cylindrical, and not in contact with each other. Tightly packed, four or five-sided corallites are, however, found in the species *L. basaltiformis*. Within the outer wall the various species are characterized by a small, rod-like, central boss. The septa are thickened and short and the tabulae conical in shape. *Lithostrotion* is found worldwide and really excellent material is found throughout Europe. In the USA various species occur in the Lower Carboniferous (Mississippian) sediments of Montana, Kentucky and Missouri.

Lonsdaelia   Although *Lonsdaelia* is superficially similar to *Lithostrotion*, sufficient features exist to distinguish between the two in the field. Both form massive, tightly packed colonies and have angular-walled corallites. *Lonsdaelia*, however, lacks a central boss. It is replaced by a large axial structure which is surrounded by a deep, circular pit. The outer walls are strong and the septa and dissepiments well developed. *Lonsdaelia* is recorded from the Carboniferous sediments of North America, Europe, Asia and Australasia.

Acervularia   This genus forms medium to massive colonies in which the individual corallites are also large (6–15 mm / $\frac{1}{4}$–$\frac{3}{5}$ in diameter). They are often four-sided, with small juvenile corallites inside the walls of the larger units. Both septa and dissepiments are well developed. The ends of the outer septa form a distinct inner rim to the corallite. *Acervularia* is known from the Silurian of Europe. *A. ananas*, the so-called 'pineapple coral', is common in the Wenlockian sediments of England.

Phillipsastraea   A colonial coral that belongs to the same family as *Zaphrentis* and *Lithostrotion*. It is also common in Lower Carboniferous (Mississippian) sediments. Dissepiments are more prominent than in *Lithostrotion* and have a distinctive, horseshoe-shape.

Hexagonaria

Lithostrotion

Lonsdaelia

Acervularia

Phillipsastraea

5 cm

Scleractinian corals (hexacorals)

Mesozoic and Cainozoic corals, solitary or colonial, with prominent septa and dissepiments. Range Triassic to Recent. Worldwide.

SOLITARY SCLERACTINIANS

Trochocyathus   This small solitary coral varies in shape from an inverted cone to a slender horn. The skeleton is enclosed within a distinct outer wall – the epitheca – and major and minor septa are well developed. A central, spongy complex – the columella – is also visible. *Trochocyathus* is well represented by several existing species that may be found in waters between 35 and 1500 m (approximately 115 and 5000 ft) in depth. The earliest specimens of this species are found in rocks of Upper Jurassic age. The genus is known worldwide, with good material collected from the Eocene of California, USA.

Turbinolia   Another small solitary coral characteristic of the Eocene of the USA (California and Alabama) and northern Europe. The skeleton is horn-shaped and the major septa thick and strongly developed. A star-shaped central columella is also characteristic of this genus. The outer surface of the skeleton appears strongly ribbed due to the development of the septa and the outer wall. *Turbinolia* ranges from the Eocene to the Recent and is known worldwide.

Montlivaltia   A relatively large, solitary coral with horn-shaped to cylindrically shaped corallites. The outer wall is rather thin and is often eroded to display the vertical septal ridges. The septa are numerous and their upper surfaces have a serrated appearance. *Montlivaltia* existed from Triassic to Upper Cretaceous times. It is worldwide in distribution, and good material is recorded from the Upper Jurassic of Germany and England, and from the Upper Triassic of the French Alps.

Microbachia   This wedge-shaped solitary coral is characterized by strong vertical ridges on the outer surface. The major septa are well developed and they almost fuse with the central columella. The minor septa are short. *Microbachia* is closely related to *Turbinolia*. Living species exist in outer shelf environments. They are non-reef dwelling i.e. ahermatypic corals. The genus is known worldwide since the Eocene. Excellent specimens are recorded from the Tertiary deposits of Europe and North America.

Parasmilia   Contemporary representatives of this small, solitary coral live at a depth of approximately 325 m (1065 ft). The skeleton tapers downwards but flares at the base into an attachment structure. Each corallite is circular in section, with numerous septa characterized by a granular surface texture. The central axial complex is large and spongy. Vertical septal ridges may be strongly developed on the outer surface. *Parasmilia* is known worldwide, from the Cretaceous.

Trochocyathus

Turbinolia

Microbachia

Montlivaltia

Parasmilia

cm

COLONIAL SCLERACTINIANS

Isastrea   A massive colonial coral composed of numerous closely packed corallites. The colonies are unbranched, and usually circular and domed in appearance. The individual corallites are thin-walled, and five or six-sided. In some species the septa cross the bounding walls to fuse with those of adjacent corallites. Numerous septa are present but *Isastrea* lacks a central columella. It ranges from the Middle Jurassic to the Cretaceous and is found in Europe, Africa and North America. Important sites occur in the Jurassic beds of Gloucestershire, England.

Thamnasteria   This massive colonial coral is often found in association with *Isastrea*. The corallum is usually branched and 'stick-like'. The individual corallites are small to medium-sized, with ill-defined boundaries; this is due to the absence of the outer wall and the fusion of adjacent septa. A slender axial structure is present. *Thamnasteria* was an important reef-builder during the Jurassic. Its range is Triassic to Cretaceous, and it is found in North America (Upper Triassic, California), Europe (Jurassic, Gloucestershire, England), South America and Asia.

Porites   This is a comparatively short-ranged genus. It forms massive, branching or encrusting colonies and is found in modern reef environments. The individual corallites are small, without well-defined outer walls. Few septa are present in each corallite, those which are present are short and rather spiny and do not join up with the axial structure. *Porites* is known worldwide and is first recorded from sediments of Eocene age.

Astrocoenia   A massive, colonial coral with medium-sized corallites. The septa are strongly developed and somewhat restricted in numbers. The outer wall is poorly defined in various species. *Astrocoenia* ranges from Upper Jurassic to Recent, and is cosmopolitan in its distribution. Excellent specimens are recorded from the Lower Cretaceous of Texas, USA, and the Tertiary of the Paris Basin, France.

Favia   This very distinctive colonial coral possesses small to medium-sized corallites. These are closely packed to form a massive, columnar or disc-shaped corallum. The individual corallites are separated by a wall of rather spongy tissue, which is traversed by outward extensions of the septa. The septa are of variable length with a limited number fusing with the large, spongy axial structure. Dissepiments are well developed in this genus. *Favia* first appeared in the Cretaceous, and is known worldwide at the present time.

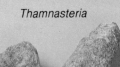

Isastrea

Thamnasteria

Porites

Astrocoenia

Favia

5 cm

Jellyfish, hydrozoans and stromatoporoids

Other representatives of the Phylum Cnidaria apart from the corals occur within the fossil record. These include the jellyfish, hydrozoans, and stromatoporoids.

The jellyfish are soft-bodied, medusoid creatures that are unknown other than as impressions, moulds or casts. They are, however, among the oldest known fossils, with animals remarkably similar to those that exist today recorded from the Precambrian Ediacara Formation of Australia. *Medusina* is a representative genus of this Precambrian fauna. The fossilization is quite unique, with the soft parts preserved as impressions on a rather coarse, sandy substrate.

In contrast the skeletons of the hydrozoans and stromatoporoids are preserved mostly as three-dimensional structures. Distant relatives of the hydrozoans are also recorded from the Ediacara Formation. The stromatoporoids are possibly related to living hydrozoans but their complex structure has made any positive association difficult to confirm. The upper surface of a stromatoporoid colony is often covered in raised areas or swellings, termed **mamelons**. These are pierced by small vertical canals which spread out into a radially arranged series of grooves, the **astrorhizae**. Some genera have a pillared structure with thin horizontal laminae or tabulae between them. *Stromatopora* is a typical stromatoporoid, and ranges from the Ordovician to the Permian. The stromatoporoids were important reef builders during the Palaeozoic. Modern hydrozoans belonging to the Milleporina (such as *Millepora*) and Stylasterina (such as *Astya*) are also important components of reef communities. They superficially resemble the corals, by virtue of their vertical tubes and cross-partitions. The skeleton is thick-walled and lacks septa.

The red 'organ pipe' coral *Tubipora* is representative of another group of cnidarians, the octocorals. These are poorly represented in the fossil record and are first recorded from Cretaceous sediments. Their skeletons consist of horny corallites that are often found in massive colonies. Small calcite spicules are set within the horny tissue. The small polyps of the octocorals have eight tentacles.

The 'sea-pens' are also octocorals. Their colonies are shrub- or feather-like, with each branch supporting a large number of polyps. The 'sea-pens' are possibly the oldest of all hydrozoans, as it is the constituent genus *Rangea* that is found in the sediments of the Ediacara Formation. The Recent genus *Pennatula* is typical of this group.

Medusina

Stromatopora
(×1)

Millepora

Astya

Pennatula

Tubipora

5 cm

Bryozoans

Delicate to massive, colonial organisms. Skeletons often plant-like in appearance, branched, and possessing small tubes or **zooecia**. Range Upper Cambrian to Recent.

Fenestella This so-called 'lace bryozoan' is common in sediments of Upper Palaeozoic age. The skeleton consists of a large number of radiating, slender branches linked by thinner cross-bars or dissepiments. Each branch is pocked by the openings of numerous zooecia. The whole skeleton is fan-shaped or funnel-like in appearance. *Fenestella* has an Ordovician to Permian range and is found in Europe. *Fenestrellina*, a related genus, is found in the Carboniferous of North America.

Meandropora This bryozoan is characterized by medium-sized to massive colonies. These are comprised of narrow rod-like tubes that branch frequently. The colony occurs in rounded or globular masses which have a rough outer surface. Excellent examples of *Meandropora* are found in the sandy crags of East Anglia in England. It is also known as *Fascicularia* in some texts, this name describing its 'bundle of rods' appearance.

Archimedes This distinctive bryozoan is distinguished by its screw-like axis. The latter is solid, and provides the support for spiralled, lace-like fronds. These fronds resemble those of *Fenestella*, but the spiral is achieved by the localized deposition of additional skeletal tissue. Two rows of openings occur on each of the branches. *Archimedes* is known from Europe, North America and Asia. It is found only in Carboniferous to Permian sediments, with the best examples occurring in the Carboniferous. Good specimens are recorded from the Lower Mississippian of western Illinois, USA.

Berenicea Usually found as an encrusting, colonial organism, *Berenicea* is a small cyclostome bryozoan. Its colonies are mostly circular in form, the individual zooecia having thick walls and rounded apertures. They are also closely packed, and appear to radiate outwards from a central nucleus. *Berenicea* ranges from Ordovician to Recent. It is commonly found encrusting Jurassic and Cretaceous shells, with numerous examples recorded from the chalk of northern Europe.

Theonoa *Theonoa* is known to occur either as erect, branching colonies or as rounded masses. The colonies are usually described as massive, with the large main tubes separated by a spongy tissue. *Theonoa* is a representative of the cyclostome bryozoans. The group is best known from Palaeozoic and early Mesozoic rocks. *Theonoa* is recorded from rocks of Jurassic and Cretaceous age. It is known throughout northern Europe, with excellent samples collected from the Farringdon Sponge Gravels of southern England.

Fenestella

5 cm

Meandropora

Archimedes

Berenicea
(×2·5)

Theonoa

Brachiopods

Exclusively marine animals in which the shell is made up of two valves of unequal size. The shell exhibits equilateral symmetry and is composed of calcite or chitinophosphatic substances. The phylum is divided into two classes, the Inarticulata and the Articulata. Range Cambrian to Recent.

Inarticulate brachiopods

The majority of inarticulate brachiopods are small with rounded, indistinct hinge lines. Teeth and sockets are absent, as are internal support structures for the feeding organ. The shell is usually chitinophosphatic, but a number of genera have calcareous shells.

Orbiculoidea A small brachiopod characterized by a chitinophosphatic shell marked by concentric growth lines; this indicates that the valves have increased in size around their complete margin. The brachial valve is conical to subconical in shape, with the apex in the central position. The pedicle valve is flatter, or even concave with a short closed pedicle track. *Orbiculoidea* is known worldwide from rocks of Ordovician to Permian age.

Lingula The modern *Lingula* lives in inshore brackish and intertidal areas. It has a chitinophosphatic shell with a characteristic elongate shape. The two unequal valves, which are marked with fine growth lines, exhibit a bilateral symmetry. They have a short, curved hinge line and are marked internally by a number of strongly defined muscle scars. *Lingula* and the related Palaeozoic genus *Lingulella* show little or no change throughout their histories. Both are common in the dark shales of the Lower Palaeozoic. *Lingula* has a range of over 500 million years, from the Cambrian to Recent, and is known worldwide.

Crania A small subcircular shell composed of calcium carbonate. There is no pedicle opening present as the shells are found cemented to other organisms. *Crania* is known worldwide, from the Cretaceous to Recent.

Acrothele This small inarticulate brachiopod has a semicircular shaped shell with a straight hinge line. The brachial valve is flat or gently convex. Both valves have an ornamentation of fine concentric growth lines. The surface texture is granular. Two short spines may be present on the apex of the beak of the brachial valve. *Acrothele* is recorded from the Middle Cambrian of Europe, North America, Asia and Australia.

Orbiculoidea

Lingula

Lingulella

Crania

Acrothele

5 cm

Articulate brachiopods

Brachiopods with a calcareous shell. The hinge line is well developed, and internal support structures are developed in many families. When the hinge line is less than the maximum width of the shell the term **non-strophic** applies. Genera in which the width of the hinge line is equal to, or greater than, the width of the shell are described as **strophic**.

ORTHIDS   Shells unequally biconvex, strophic and with ribbed ornament.

Heterorthis   Shell roughly subcircular in shape. It has a convex pedicle valve, while the brachial valve is either concave or flat. The valves are finely ornamented with radial ribbing and concentric growth lines. *Heterorthis* is known only from the Middle and Upper Ordovician sediments of Europe and North America.

Dinorthis   A Middle to Upper Ordovician orthid with a rounded outline and a wide straight hinge line. The shell has a convex pedicle valve and a concave brachial valve. Externally the shell is marked by a strong ribbing. *Dinorthis* is common in the Middle Ordovician sediments of Europe and North America.

Platystrophia   This genus is unusually large for an orthid. It is also biconvex with a strong beak that curves inwards. A sharp fold occurs on the brachial valve; this is matched by a sulcus on the pedicle valve. *Platystrophia* has a strongly ribbed ornament crossed by numerous concentric growth lines. The genus bears a close resemblance to some spiriferids. Known worldwide, *Platystrophia* is found in sediments of Ordovician and Silurian age in Europe and the USA.

Dalmanella   A small to medium-sized orthid with a distinct, almost circular shape. The shell is biconvex, with the brachial valve more convex than the pedicle valve. The valves have a ribbed ornamentation in which the ribs are of variable thickness. Strong concentric growth lines are well developed near the margins of the valves. *Dalmanella* is known worldwide from the Ordovician and Silurian.

Dolerorthis   *Dolerorthis* is closely related to *Dalmanella*. It has a charac-teristic rounded outline and a strongly ribbed ornament. The hinge line is long, with a triangular opening for the pedicle. Prominent muscle scars occur internally. *Dolerorthis* is confined to Lower and Middle Silurian sediments of Europe and North America.

Dicoelesia   This unusual orthid is characterized by a deep indentation on the front edge of the shell and a non-strophic hinge line. The valves are sometimes smooth but frequently ornamented with fine ridges and concentric growth lines. *Dicoelesia* has an Ordovician to Devonian range and is known worldwide.

Heterorthis

Dinorthis

Platystrophia

Dalmanella

5 cm

Dolerorthis

Dicoelesia

RHYNCHONELLIDS Small to medium-sized brachiopods. Biconvex and rounded, with prominent beak and coarsely ribbed ornamentation.

Tetrarhynchia A small to medium-sized brachiopod which is broader than it is long. It has a subtriangular shape and a fairly small, curved beak. It is coarsely ribbed, although the ribbing is less distinct towards the hinge line. A strong fold is present in the brachial valve, with a corresponding flexure in the pedicle valve. *Tetrarhynchia* is recorded from the Jurassic of Europe and North America.

Camarotoechia This well-known rhynchonellid has a small shell characterized by its suboval shape. The hinge line is rounded and non-strophic. A small incurved beak is present on the pedicle valve. The shell has a distinct fold in the brachial valve and a corresponding sulcus in the ventral, pedicle valve. *Camarotoechia* is recorded from the Silurian and Devonian of Europe and North America.

Rhactorhynchia This is a medium to large-sized rhynchonellid. It is characteristically rounded and swollen in appearance, and strongly ribbed. The pedicle valve has a strong beak and a small pedicle opening. Both valves are convex. The anterior opening of the shell is noted for the presence of a poorly developed and asymmetric fold. *Rhactorhynchia* is confined to the Jurassic and is recorded worldwide. It was a common component of shallow-water communities in Europe and North America.

Sphaerirhynchia A small to medium-sized rhynchonellid known essentially from the Middle Silurian of north-west Europe. It has a globose, subrounded to cubic appearance with the front and sides squared off. The anterior opening – the commissure – has the form of a broad, box-like fold. Both valves are coarsely ribbed and the pedicle valve has a small but well-developed beak. In weathered specimens it is possible to recognize a pronounced septum that forms part of the internal skeleton. *Sphaerirhynchia* is also recorded from Devonian sediments in Oklahoma and New York State, USA.

Rhynchonella This is the type genus for the order, the first specimen having been named in 1809. *Rhynchonella* is small to medium-sized and triangular in shape. It has a smoother shell than most rhynchonellids, with only a few large ribs (costae) at the front of the shell. The shell is inflated with a strong anterior fold. The genus is recorded from the Jurassic of Europe, and is particularly common in limestones and muddy limestones.

Cyclothyris A small to medium-sized rhynchonellid with a prominent beak. It has many costae that radiate from the hinge line. The hinge line is non-strophic. A well-developed pedicle opening is present. The anterior margin of the shell is gently folded. The shell is wider than it is long. *Cyclothyris* is known from the Lower Cretaceous of Europe and North America.

Tetrarhynchia

Camarotoechia

Rhactorhynchia

Sphaerirhynchia

5 cm

Cyclothyris

Rhynchonella

PENTAMERIDS   Medium to large-sized brachiopods. Biconvex with short, rounded hinge line. In section, the pentamerids are identified by the presence of organ support and muscle attachment structures. These 'hang' from the brachial valve or branch upwards from the pedicle valve. The **brachial process** supports the feeding organ, whereas the **spongylium**, in the pedicle valve, is the seat for muscle attachment. Both structures reach towards the centre of the shell and form a diamond-like structure.

Conchidium   A large brachiopod with a markedly biconvex shell. The pedicle valve is larger than the brachial valve. It possesses a large beak that curves upwards to enclose the beak of the brachial valve. Both valves are strongly ribbed. The anterior opening is straight. *Conchidium* is known from Silurian and Devonian rocks. It is distributed throughout the world and is particularly abundant in shallow-water limestone deposits.

Pentamerus   One of the largest of the pentamerid brachiopods, *Pentamerus* has a smooth, rather elongate shell. The shell is markedly biconvex, with the pedicle valve having a prominent beak and being the larger of the two valves. Internally the shell is divided by a well-developed internal support structure. This often results in the shell splitting during collection. *Pentamerus* is known only from the Silurian. It is a common fossil in the Welsh borders, Great Britain, and in Illinois, USA.

Gypidula   This small to medium-sized pentamerid is a common fossil in Silurian and Devonian limestones. The shell is biconvex and elongate to subcircular in shape. It may be marked by the presence of strong ribs and concentric growth lines. The anterior opening is folded and more crenulate than those of *Pentamerus* or *Conchidium*. A large beak is characteristic of the pedicle valve. *Gypidula* is known from the Northern Hemisphere during Silurian and Devonian periods. It is particularly common in the Devonian of the eastern states of the USA.

Stricklandia   A large pentamerid with a smooth, subcircular to elongate shape. The shell is gently folded anteriorly. It is moderately convex. The internal support structure is fairly large. *Costistricklandia*, a related genus, is strongly ribbed. Both are recorded from the Silurian of Europe and North America. *Stricklandia* is known from the Llandovery Beds of the Welsh borders, Great Britain, and the Appalachian area of North America.

Conchidium

Gypidula

Pentamerus

Stricklandia

5 cm

TEREBRATULIDS   Biconvex, non-strophic brachiopods. Rounded to elongate in shape with a strongly developed umbo, and pedicle opening on pedicle valve. Shells usually smooth, rarely ribbed. Internal skeleton in the form of a loop. Range Lower Devonian to Recent.

Gibbithyris   This brachiopod has a small, smooth, strongly biconvex shell. It may be termed inflated, and has a somewhat rounded appearance. The surface is marked by well-developed and closely packed growth lines. A small, but well-developed pedicle opening is present on the pedicle valve. The shell is broadly folded anteriorly. *Gibbithyris* is known from the Upper Cretaceous of England.

Terebratula   This is the type genus on which the Order Terebratulida was erected. It was first named in 1776. *Terebratula* is a medium to large-sized brachiopod with a biconvex shell. The shell is smooth with concentric growth lines. It is flattened towards the anterior of the valves. The shell is broadly pear-shaped with a prominent pedicle opening. Internally, the hinge teeth have swollen bases and the internal skeleton or loop is triangular in shape. *Terebratula* is known from the Miocene and Pliocene of Europe, although related genera may be traced back to the Triassic and are worldwide in their distribution.

Dielasma   This genus is the typical representative of this family of small to large-sized terebratulids. It has a smooth shell with a broad dorsal fold and ventral sulcus. The brachial valve has a prominent, incurved beak. Concentric growth lines occur on both valves. The internal support is in the form of a short loop. *Dielasma* is known worldwide from Carboniferous and Permian deposits.

Dictyothyris   A small to medium-sized terebratulid with a distinctive shell. The anterior is W-shaped. A delicate ornament of radiating striae (raised lines) and concentric growth lines is characteristic. These intersect to give a 'spiny' reticulate patterning. The umbo is short and curved. *Dictyothyris* is recorded from the Middle and Upper Jurassic of Europe.

Stringocephalus   A large, thick-shelled terebratulid with a subglobular or sublenticular shape. The beak is large and sharp. A large triangular pedicle opening is present. The internal skeleton consists of a broad loop with inwardly directed spine line projections. *Stringocephalus* is a rather unique fossil. It is confined to the Middle Devonian sediments of Europe, North America and Asia.

Sellithyris   *Sellithyris* is known from England, France, Belgium, Germany and Switzerland. It is recorded from sediments of Cretaceous age and is particularly common in the Isle of Wight, England. It is a small to medium-sized terebratulid with a smooth, moderately biconvex shell. The anterior is characterized by two closely spaced folds and is termed biplicate. A rounded pedicle foramen is present at the tip of the umbo.

Gibbithyris

Terebratula

Dielasma

Dictyothyris

5 cm

Stringocephalus

Sellithyris

STROPHOMENIDS   A variable and long-ranging group of articulate brachiopods. The shells are strophic, with a wide hinge line. They may be **plano-convex** to **concavo-convex** in profile, with a few biconvex genera also recorded. The pedicle opening is closed in most forms. Several suborders may be identified on the presence of ribs, spines and other diagnostic characters.

Sowerbyella   *Sowerbyella* has a long straight hinge line and a fine ornament of radial striae. The shell is semicircular in shape. The genus is often found as internal casts that show two large ridges for the attachment of muscles. *Sowerbyella* has a concave brachial valve and a convex pedicle valve. It is cosmopolitan in distribution during the Ordovician and Lower Silurian.

Kjaerina   A medium-sized brachiopod in which the brachial valve is concave and the pedicle valve convex. The shell is semicircular in shape with distinct growth lines and radial costae. The latter are less pronounced towards the hinge line. A raised mid-line ridge is characteristic. *Kjaerina* is related to the genus *Rafinesquina*. Both are known from the Middle and Upper Ordovician of Europe and North America. *Kjaerina* is common in the Middle Ordovician sediments of New York State, USA.

Stropheodonta (Strophodonta)   Like *Leptaena* and *Sowerbyella*, this genus has a concavo-convex shell. The shell is longer than it is wide but again has a semicircular shape. The hinge line is strophic and marked internally by a series of small perpendicular grooves. *Stropheodonta* has a strong radial ornamentation. It is known worldwide from the Upper Ordovician to the Upper Devonian.

Leptaena   Probably one of the best known of all strophomenids, *Leptaena* has a concave brachial valve and a convex pedicle valve. Anteriorly the shell bends at a sharp angle to give an L-shaped profile. It is marked by fine radial striations and strong concentric ridges (rugae). *L. rhomboidalis* is known from the Upper Silurian of the Welsh borders, Great Britain, and probably from the Middle Devonian of Michigan, USA. The genus *Leptaena* is known worldwide. It has a Middle Ordovician to Devonian range.

Productus   This genus is a representative of an extremely diverse group of strophomenids. The productids include the largest and the most bizarre of all brachiopods. *Gigantoproductus* may have a width of over 15 cm (6 in). *Productus* itself is a small to medium-sized brachiopod with a hemispherical shape. The pedicle valve is strongly convex, the brachial valve concave or lid-like. *Productus* may be marked externally by faint ribs, pustules or with spines. It is recorded from the Carboniferous sediments of Europe and Asia. Associated genera are well known in North America (e.g. *Linoproductus* from the Pennsylvanian [Upper Carboniferous] of Oklahoma).

Sowerbyella

Kjaerina

Stropheodonta

Leptaena

5 cm

Productus

SPIRIFERIDS These are unusual brachiopods in that shells are often very wide with wing-like extensions. They are strophic, and usually biconvex with strong ribbing and deep folds. The internal skeleton is usually in the form of a pair of spirally coiled supports that extend laterally into each wing. Early genera, such as *Meristina*, are rounded in shape. Range Middle Ordovician to Jurassic.

Meristina An early spiriferid with a biconvex shell that is longer than it is wide. The beak on the pedicle valve is well developed and uncurved in larger, older specimens. Both valves are smooth with only faint growth lines. The anterior of the shell is gently folded. *Meristina* is common in Silurian and Devonian sediments. Excellent samples can be collected in the Silurian limestones of the Ludlow region of Shropshire, England, and from the Upper Silurian of Indiana, USA.

Atrypa Like *Meristina* this genus is also an early and rather untypical spiriferid. It has a biconvex, although somewhat flattened, shell that is subcircular in shape. The beak is small and the hinge line non-strophic. *Atrypa* has a strong ornamentation of concentric ridges cross-cutting numerous radiating striae. The anterior area is gently flexed. *Atrypa* has a Silurian to Devonian range and is known worldwide.

Mucrospirifer *Mucrospirifer* is a beautiful example of a 'winged' or broad-hinged spiriferid. The hinge line is very wide and the body of the shell is seen to taper sharply on each side. The central region is biconvex with a strong fold and sulcus. The ornament consists of strong ridges. *Mucrospirifer* is confined to the Devonian but is known worldwide.

Cyrtospirifer A biconvex, medium-sized brachiopod. The hinge line is equal to the maximum width of the shell. A strong fold and sulcus occur anteriorly, with the sulcus tapering toward the hinge line. The shell has an ornamentation of radial striae. A well-developed incurved beak is present in mature specimens. The genus is recorded from Upper Devonian and Lower Carboniferous (Mississippian) sediments. It is abundant in the Devonian limestones of Belgium.

Spirifer A biconvex brachiopod with a wide hinge line. The shell is semicircular to semi-elliptical in shape. It has a gentle fold anteriorly. The umbo is well developed and incurved. The external ornament consists of radiating ridges and concentric growth lamellae. The internal skeleton is in the form of a pair of spirally coiled supports (spiralia). *Spirifer* is known only from the Carboniferous. It is distributed worldwide.

Composita This small to medium-sized spiriferid has a biconvex shell. Superficially it resembles a terebratulid, as the shell is longer than it is wide and is smooth. A dorsal fold and ventral sulcus are evident. The pedicle opening is well developed. *Composita* is recorded from the Upper Palaeozoic and is known from Europe, North America and Australia.

Meristina

Atrypa

Cyrtospirifer

Mucrospirifer

5 cm

Spirifer

Composita

Molluscs

A group of invertebrates which includes the chitons, gastropods, scaphopods, bivalves and cephalopods. Range Cambrian to Recent.

Gastropods

Univalved molluscs in which the shell is usually coiled either **planispirally** or **helically**. The shell is not chambered. Cambrian to Recent.

ARCHAEOGASTROPODS These gastropods have either planispiral or helically coiled shells. They have no siphonal notch or canal, but a slit is frequent. Range Lower Cambrian to Recent.

Bellerophon A small to large sized gastropod with a wide shell. The shell is bilaterally symmetrical with a deep slit on the front margin of the opening (aperture). In most specimens the slit is gradually infilled and may be represented as a distinct ridge. The anterior end of the shell is flared. *Bellerophon* is known worldwide from the Silurian to the Triassic.

Poleumita *Poleumita* is a coiled gastropod with a flattened upper surface. The whorls have a pronounced, knobbly ridge (shoulder) in the centre of the coils. A fine ornament of radial lamellae is present. The aperture varies in shape, but is usually angular in appearance. *Poleumita* is confined to the Silurian sediments of Europe and North America.

Pleurotomaria *Pleurotomaria* has a conical spiral shell with a wide apical angle. The aperture is large with a somewhat flattened appearance in some shells. A prominent slit occurs on the outer lip. The whorls increase gradually towards the aperture and the shell tends to be slightly asymmetric in shape. *Pleurotomaria* occurs in Jurassic and Cretaceous sediments, worldwide.

Euomphalus This well-known genus is similar to *Strapolus*, a related form from Devonian to Permian strata. *Euomphalus* has a rather delicate lined ornament that flexes over the raised shoulder of each whorl. The genus has a Silurian to Permian range throughout Europe and North America.

Trochus *Trochus* is a medium-sized gastropod with a conical shell in which the apical angle is small. The aperture is situated more or less ventrally and the inner lip appears thickened. A pronounced granulose ornamentation is characteristic. The whorls are flat-sided. *Trochus* is known worldwide from Miocene to Recent.

Diodora A small to medium-sized gastropod with a conical shell. It is limpet-shaped but with a small opening near the apex. The shell has an ornament of coarse radiating ribs crossed by concentric growth lines. This type of ornament is termed cancellate. *Diodora* first appeared in the Upper Cretaceous. It is recorded from North America, South America, Africa and Australasia.

Bellerophon

Poleumita

Pleurotomaria

Euomphalus

5 cm

Trochus

Diodora

MESOGASTROPODS   The shells of this major group of gastropods vary considerably in shape and size. The aperture of many species has a prominent **siphonal canal**. Shells may be smooth or well ornamented. Range Ordovician to Recent.

Cerithium   This has a tall turreted shell with a narrow apical angle. There are many whorls, each with a distinct shoulder. The ornament consists of vertical ridges and spiral growth lines. The aperture is relatively small and notched. *Cerithium* is abundant in the Tertiary rocks of the Paris Basin, France, and Texas, USA. It is known worldwide and has a Cretaceous to Recent range.

Turritella   This genus, as its name suggests, has a long turreted spire. The apical angle is acute. The spire is formed of numerous whorls, the side walls of which are either flattened or slightly rounded. Dependent on species, the ornament may be smooth or in the form of spiral ridges and the normal growth striations. The body whorl is only slightly larger than the one before and the aperture is entire. *Turritella* is known from the Eocene onwards and is distributed worldwide.

Strombus   A moderately large form with a turreted spire. The aperture is long and the lips tend to flare in adult specimens. *Strombus* is a shallow-marine herbivore that feeds on red seaweeds. It is found today on the beaches of Florida, USA, and Africa. It has a colourful shell marked by both ribs and notches. *Strombus* first appeared during the Cretaceous, and is cosmopolitan in its distribution.

Aporrhais   *Aporrhais* is related to *Strombus* and has a turreted spire. It is distinguished by the presence of a flared, often spinose, wing that extends away from the outer lip. The apical angle is narrow and the whorls rounded in outline. A bold sculpture of vertical ridges and strongly developed tubercles may be present. Fine growth lines can be observed on the outer lip and wing. *Aporrhais* is a shallow burrower. It is first recorded from Jurassic sediments and lives today in the North Atlantic. In past times it was distributed worldwide.

Natica   This shell-drilling, predatory gastropod has a globular to conical shell with a wide apical angle. It is small to medium-sized, with the body whorl enormous by comparison with the others. The shell is usually smooth and the aperture entire. *Natica* has a Triassic to Recent range and is found worldwide.

Cypraea   This is a small to large-sized gastropod with a characteristic conch-shaped shell. The outer or last whorl is greatly expanded and wrapped around the inner whorls. The aperture is long and the thickened outer lip is often serrated. *Cypraea* has a smooth, shiny shell. It first appeared during the Cretaceous and exists today throughout the world. *Cypraea* and its relatives are known colloquially as the 'cowrie-shells'.

Cerithium

Turritella

Strombus

Aporrhais

5 cm

Natica

Cypraea

NEOGASTROPODS   Gastropods with a variety of shell types, in which the siphonal canal is often a very prominent feature. Range Cretaceous to Recent.

Rimella   A small to medium-sized gastropod with a spindle-shaped (fusiform) shell. The spire is elongate and the apical angle narrow. A posterior canal extends from the aperture vertically towards the apex. It adheres to the whorls and expands or opens ventrally into the inner and outer lips. The aperture is narrow and both lips are thickened. *Rimella* is strongly ornamented with vertical ribs and spiral striations. It ranges worldwide from Eocene to present day shallow-marine environments.

Athleta (Voluta)   Well known from Tertiary strata, *Athleta* is a thick-shelled, short-spired gastropod. The body whorl is very large, with an elongate siphonostromatous aperture. Short vertical spines or strong nodes occur along the shoulder of each whorl. These continue over the whorl as strong, vertical ribs. *Athleta* has a narrow aperture. The inner lip has a thin callus and a distinctly notched lower area. The genus ranges from the Tertiary to Recent and is known worldwide.

Conus   A small to large-sized gastropod with a flat to conical spire. The apical angle is large. A large body whorl is characterized by an elongate, narrow aperture which is parallel-sided. There is a distinct notch at the upper end. The ornament varies but is usually restricted to spiral grooves and delicate ridges. *Conus* first appeared in the Upper Cretaceous. *Conus* is known worldwide and persists in shallow marine environments.

Murex   *Murex* is a medium to large-sized gastropod in which there are three strong axial ribs on each whorl. Prominent spines are also present, as are spiral ridges. The aperture is small with a well developed siphonal canal. The outer lip is expanded into a spinose rib. *Murex* is known worldwide and ranges from the Cretaceous to the present day.

Voluta   Many species previously referred to the genus *Voluta* have subsequently been renamed and associated with the named genus *Athleta*. Caution should therefore be taken in the naming of your material and reference made to books such as the British Museum (Natural History) publication on *Caenozoic Fossils*. The term 'voluted' is still used, however, for many shells having nodes or spines along the shoulders and strong vertical ridges on the body whorl. The aperture is elongate with a long siphonal canal. The volutids first appear in the Cretaceous and are important constituents of many Tertiary faunas.

Neptunea   This robust form has a large body whorl and an elongate spire. The whorls are rounded and the shell is essentially smooth. Each whorl has a sloping shoulder and the sutures between whorls are strongly indented. The aperture is oval in shape with a short siphonal canal and a thickened inner lip. It is found worldwide since the Eocene.

5 cm

Rimella

Athleta

Conus

Murex

Voluta

Neptunea

PULMONATES   Land-dwelling and freshwater snails and slugs. The shells vary from moderately elevated cones with flattened or rounded bases, to discoidal shapes with the aperture at the side. The aperture is rounded and entire. Most pulmonate shells are smooth.

Pupa   *Pupa* is a rather unusual pulmonate as the shell closely resembles the pupal stage of certain insects. It is thus described as pupaeform. The length of the shell may vary from rounded to elongate. The whorls are rounded and the sutures distinct. A rounded, thick-lipped aperture occurs in a central, or slightly off-centre position. *Pupa* ranges from the Eocene to the present day. It is known worldwide.

Planorbis   A small to medium-sized shell which is flattened and discoidal in shape. It is slightly concave on the ventral surface. The sutures may be deeply impressed and the outer whorls are higher than inner ones. An oval to wide crescent-shaped aperture is characteristic of this genus. *Planorbis* first appeared in the Oligocene and is found in Europe, Africa and Asia.

Hydrobia   This small to medium-sized pulmonate has a moderately high-spired shell. It is rather similar to *Lymnaea* but the aperture is more rounded and the lower whorls broader. The apical angle is quite narrow and the ornament restricted to fine growth lines. *Hydrobia* is first found in Jurassic sediments. Its shells are sometimes found lining burrows in lake sediments. *Hydrobia* is a cosmopolitan genus.

Lymnaea   Like *Hydrobia*, this genus is a good indicator of freshwater environments and it is often found in freshwater limestones or mudstones. It is moderately spired or high-spired with an elongate oval aperture. The body whorl is long and slightly asymmetric. A thickened inner lip is also characteristic. *Lymnaea* ranges from Jurassic to Recent and is cosmopolitan in its distribution.

Physa   A left-handed shell (one with the aperture on the left) is characteristic of this genus. The spire is low to moderately elevated and the body whorl much larger than any other. The aperture is elongate and oval with both lips flattened on the outside. The shell may be smooth or, in 'shouldered' forms, axially ribbed. *Physa* has a Jurassic to Recent range and is known worldwide.

Helix   Commonly thought of as the garden snail, *Helix* has existed since the Eocene. It has a moderate to large-sized shell which is helically coiled. The apical angle is wide and the body whorl greatly inflated. *Helix* has a rounded aperture bordered by a thin outer lip. The shell ornament is usually smooth although some species exhibit growth lamellae. *H. ramandii* is a well-known species from the Oligocene marls of central France. The genus is known worldwide.

Pupa

Planorbis

Hydrobia
(×17·5)

Lymnaea

Physa

5 cm

Helix

Bivalves

Marine or freshwater animals in which the shell consists of two **valves**.

PALAEOTAXODONTS Small bivalves in which the hinge line bears many small, subequal teeth. Mostly shallow burrowers. Range Ordovician to Recent.

Nucula A small bivalve in which the valves are of equal size and the anterior and posterior areas are approximately of equal length. The beaks point posteriorly. Both valves may possess a fine ribbed ornamentation and distinct growth lines. The hinge line has numerous subequal teeth. Range Upper Cretaceous to Recent, worldwide.

Acila The shells of *Acila* are usually quadrangular, but both are ovate in shape. The shell ornamentation consists of widely spaced growth lines and pairs of faint ridges. Numerous subequal teeth occur along the hinge line. *Acila* is a burrowing bivalve. It has a Cretaceous to Recent range and is known from Europe, the Americas, Asia and North Africa.

PTERIOMORPHS Members of this group are variable in shape and size and have different teeth arrangements. Range Ordovician to Recent.

Parallelodon A large bivalve with an elongate, biconvex shell. The valves are twice as long as they are high and the front end is much shortened. The hinge line is straight and has two clusters of teeth. Externally the shell is marked by strong concentric growth lines and a radial ribbing. *Parallelodon* ranges from the Devonian to the end of the Jurassic, and is known worldwide.

Arca This medium-sized taxodont bivalve has an elongate shell. The hinge line is straight with wide flattened areas separating the beaks. A large number of comb-like teeth occur along the hinge. The lower margin of the shell is often characterized by a wide gape. Both valves are convex with an ornamentation of radial ribs and concentric growth lines. *Arca* first appeared in the Jurassic and exists worldwide today.

Glycimeris This small to medium-sized bivalve is almost circular in outline. The beak is placed centrally and the surface is finely ornamented with radial ribs and concentric growth lines. Numerous teeth occur along the slightly curved hinge line. *Glycimeris* ranges from Cretaceous to Recent and is found worldwide.

Mytilus This has a shell composed of two equal-sized valves which are elongate and fixed to rocks. The shell is slipper-shaped; externally it exhibits well-defined concentric growth lines. The shell is usually toothless and often quite thin. The genus first appeared in the Triassic and occurs worldwide today.

5 cm

Acila

Nucula

Arca

Parallelodon

Mytilus

Glycimeris

Gryphaea This bivalve has valves of unequal proportions. The left, lower valve is deeply convex with a large beak. The valves have a layered appearance due to the deposition of successive growth lamellae. A large muscle scar occurs on the inner surfaces of the two valves. *Gryphaea* is found worldwide in Upper Triassic and Lower Cretaceous rocks.

Gervillella The shell of *Gervillella* is elongate with a sharply pointed front; it has an ornamentation of concentric growth lamellae. *Gervillella* is cosmopolitan, but is known only from the Mesozoic.

Pinna A medium to large-sized bivalve with a tooth-shaped shell. The valves are of equal size and may be smooth or marked by concentric growth lines and long ridges. They taper towards the anterior, and posteriorly are separated by a wide gape. *Pinna* first appeared in the Lower Carboniferous and is found worldwide today.

Inoceramus This is found in abundance in the chalk of northern Europe. It is a medium to large-sized bivalve with a strong ornamentation of concentric growth lines and ripple-like structures. A back wing is found in many specimens. *Inoceramus* is found worldwide in rocks of Jurassic and Cretaceous age.

Ostrea The shells of oysters are flat and generally orb-shaped. Both valves are convex and have a layered appearance. Concentric growth lines and strong radial ribs may be present in individual species. A large rounded muscle scar occurs in each valve. The oysters first appeared in the Cretaceous, and are distributed worldwide.

Pterinea The valves are of different sizes and have a layered appearance outside. The shells are subrounded in shape with a distinct 'ear' and wing extensions to the hinge line. Concentric growth lines are the prominent markings on the outside of the shell. *Pterinea* is known worldwide from the Upper Ordovician to the Lower Devonian.

Plagiostoma (Lima) A medium to large-sized bivalve in which the valves are of equal size. The beak points backward while the hinge area is angular with a small ear-like extension anteriorly. Externally the shell is often smooth but radial lines and weak ribs may ocur in some species. A single large muscle scar is present in the centre of each valve. *Plagiostoma* has a worldwide Middle Triassic to Upper Cretaceous range.

Pecten This small to medium-sized dysodont is characterized by the presence of anterior and posterior wings along the hinge line. The valves are almost symmetrical with strong radially arranged folds on the outer surface. *Pecten* has no evidence of teeth along the hinge but a well-developed triangular notch for the ligament is present in each valve. *Pecten* is first found in sediments of Upper Eocene age. The genus is found worldwide.

Gryphaea

5 cm

Gervillella

Pinna

Inoceramus

Ostrea

Pterinea

Plagiostoma

Pecten

PALAEOHETERODONTS   Bivalves in which the cardinal and lateral teeth are not fully differentiated from one another. Tooth number may be reduced to two very large, divergent teeth as in *Trigonia*. Shell shapes vary. Range Ordovician to Recent.

Carbonicola   This genus has equal-sized valves that are subtriangular in shape. The hinge line is curved and the external markings are restricted to strong, concentric growth lines. *Carbonicola* is known only from the Upper Carboniferous coal measures.

Modiolopsis   In this genus the shell is elongate, slightly oval in shape and lacks a strong ornamentation. Internally the muscle scars are of unequal size, with the posterior the larger of the two. *Modiolopsis* is confined to Middle and Upper Ordovician sediments but is known worldwide.

Trigonia   A medium-sized bivalve with an almost triangular shell. The anterior area is much shortened. A prominent beak that points upwards and slightly backwards is present. The shell is equivalve. *Trigonia* is strongly ornamented with either concentric ridges or nodose radially directed ribs. Internally two prominent teeth with grooved surfaces occur on the hinge of the right valve. *Trigonia* is known worldwide, and ranges from the Middle Triassic to the end of the Cretaceous.

Unio   An equivalve bivalve in which the hinge teeth consist of two cardinals and two posterior teeth in the left valve. The valves are elongate, with a relatively straight hinge. Externally the shell is marked by concentric growth lines. The internal muscles are of equal to subequal size. *Unio* is known since the Triassic, and is associated with freshwater sediments. It exists today in Europe, Asia, Africa and the USSR.

HETERODONTS   These bivalves are extremely variable in shape and structure. Their teeth are often differentiated into cardinals and laterals but may be modified and reduced. Range Triassic to Recent.

Venericardia   A medium to large-sized bivalve. The valves are of equal size and trigonal in shape. Strong radiating ribs, and prominent concentric growth lines that occur in a band paralleling the margin, are characteristic. The internal muscles are of unequal size. Large cardinal teeth are present along the hinge line. The genus is known from Europe, Africa and North America during the Palaeocene and Eocene.

Chama   A small to medium-sized bivalve in which the valves are unequal in size. Externally they have a rough, layered appearance due to growth lamellae and flattened spines. Internally the muscles are subequal in size. The hinge line may be characterized by one or two cardinal teeth and several weak lateral teeth. *Chama* is recorded in Europe and America, and ranges from Palaeocene to Recent.

5 cm

Modiolopsis

Carbonicola

Trigonia

Unio

Venericardia

Chama

Hippopodium   This is a unique bivalve as it is the type and only genus referred to in the family Hippopodiidae. It has a large, inflated, strongly biconvex shell. The valves are equal in size, and have a layered surface texture. Internally, the curved hinge line is noted for the presence of two cardinal teeth (one in each valve) and a posterior lateral tooth in the right valve. *Hippopodium* is confined to the Jurassic of Europe and East Africa.

Corbula   *Corbula* is a small to medium-sized bivalve. It has a sturdy shell characterized by a well-developed posterior rostrum. The left valve is slightly smaller than the right valve. Strong ribs and thickened concentric ridges may occur in different species. Internally the muscle scars are subequal in size. The hinge teeth are reduced. *Corbula* is cosmopolitan in its distribution and ranges from the Upper Jurassic to Recent.

Crassatella   The shell of *Crassatella* is triangular or subtrapezoidal in shape. The valves are of equal size and ornamented by well-developed concentric ridges. Internally the cardinal and lateral teeth are well developed, as is the internal ligament pit. The posterior slope of the shell is quite pronounced. *Crassatella* ranges from Mid-Cretaceous to Miocene. It is known only from Europe and North America. Excellent material may be collected from the Tertiary sediments of south-east England, the Paris Basin, France, and Texas, USA.

Pholas   *Pholas* bores into soft rocks. It has an elongate to rounded shell with a slit-like gape for the pedicle. A prominent beak may occur anteriorly. The anterior area of the shell is marked by the presence of strong concentric ridges that are often pitted or toothed in character. Evidence for the presence of siphons exists with the strong sinus developed in the posterior region of the pallial line. *Pholas* is first recorded from Cretaceous rocks. It occurs today in both the Atlantic and Pacific oceans.

Teredo   Commonly known as the 'shipworm', *Teredo* is often recognized by the presence of its long, calcite-lined burrows. The shell is much reduced and effectively covers only the anterior tip of the worm-like body. Externally the ornament of the shell is variable although short, strong spines are often present and are used during the process of boring. *Teredo* first appeared during the early Eocene. It is particularly well known from the London Clay of Sheppey, south-east England. The 'shipworm' is known worldwide at the present.

Solen   This genus is representative of a family of elongate bivalves. The valves gape at both ends and are usually smooth with concentric growth lines. The beaks are placed at the anterior end and the hinge line is weak. One to three cardinal teeth may occur but the lateral teeth are reduced or absent. *Solen* itself first appeared in the Eocene, but the family to which it belongs may be traced back into the Cretaceous. It is known from Europe, North America and the Pacific region.

Corbula
(×1·8)

Crassatella

Hippopodium

5 cm

Pholas

Teredo

Solen

PACHYDONTS   The shells of these bivalves feature a limited number of heavy, blunt teeth on the hinge line. The shells are often large, and the valves of unequal size. The pachydonts are essentially modified heterodonts.

Durania   A medium to large-sized bivalve in which the lower, right valve is coral-shaped and the upper, left valve lid-like. The wall of the lower valve is very thick and marked by the presence of radial furrows. The upper valve is flattened or arched and also furrowed. Strong vertical ribs are also present on the lower valve. *Durania* is known only from Cretaceous rocks. It is recorded in Europe, North Africa, Asia and the Americas. It is particularly common in limestones of the ancient Tethys Ocean.

Hippurites   As with *Durania*, *Hippurites* is known only from Cretaceous sediments, and the lower, right valve is again cylindrical and coral-like. The outer wall of the lower valve is very thick with a regular box-like or cellular structure. The upper valve is again lid-like. Individuals can reach a very large size. *Hippurites* and *Durania* were reef-forming bivalves. Excellent reefs occur near Lisbon in Portugal and Sorrento in Italy.

ANOMALODESMATES   Bivalves with modified and reduced shells. Teeth absent or reduced. Includes a variety of boring and burrowing species. Range Triassic to Recent.

Thracia   A small to medium-sized bivalve with a smooth to granular surface. The shell may be oblong or trigonal in shape with a rounded anterior region. The valves are approximately the same size. Internally the scars are subequal. A pallial sinus is also present. There are no hinge teeth present in this genus. *Thracia* ranges from the Jurassic to Recent and is widely distributed.

Pholadomya   The shell of this genus varies in shape from trigonal to ovate. It is medium to large-sized and strongly inflated, with two biconvex valves. The valves are of equal size with the anterior area much shortened. A strong gape may be present posteriorly. *Pholadomya* has a strong ornament of radial ribs, crossed by concentric growth lines. It is a shallow, burrowing animal. Various species have existed since the Upper Triassic. The genus is known worldwide.

Durania

Hippurites

Thracia

Pholadomya

5 cm

Scaphopods, chitons and tentaculitids

The major groups of fossil molluscs are the bivalves, gastropods and cephalopods. Numerous genera and species of these classes exist within the fossil record, illustrating the fact that the Mollusca is one of the most successful and varied of invertebrate phyla. Other Mollusca are recorded, however, and occasionally a chance discovery may prove to be of great importance in extending our knowledge of a specific subclass or order.

Scaphopods and chitons are among the less well known of molluscan groups. Both are known from Upper Cambrian and Lower Ordovician sediments. The scaphopods or 'tooth shells' are marine molluscs. They are characterized by small to medium-sized shells of a rounded cylindrical nature. The shells taper posteriorly and are gently curved. They are univalved. The majority of scaphopods live on the continental shelf and their remains are often associated with outer shelf calcareous mudstones. *Dentalium* is a living representative of the 'tooth shells'. It is first recorded from sediments of Ordovician age and it is assumed that its mode of life has remained unchanged. *Dentalium* and other living scaphopods are burrowers in soft sediment.

The fossil record of the chitons is poorer than that of the 'tooth shells'. Individual plates or specimens are known since the Upper Cambrian. The chiton has a bilaterally symmetrical shell composed of seven or eight calcareous plates. Chitons live as grazers on rocks. They are abundant in shallow-water tropical areas. A large ventral foot attaches the animal firmly to its chosen substrate.

Another group of rare or little-known molluscs are the tentaculitids. In truth their zoological affinities are doubtful, but their straight calcareous cones are generally thought to be linked with the scaphopods and related Mollusca. The shells are univalved and many species feature circular ridges. The tentaculitids are often found in great numbers and they are usually associated with shallow-water sediments such as calcareous sandstones or calc-arenites. It is likely that they rested on or partially within the sediment on the sea floor. The tentaculitids are known worldwide and are particularly common in Lower and Middle Palaeozoic (Ordovician and Devonian) sediments.

Other associated groups of elongate univalved molluscs occur in the Palaeozoic. These include the conularids in which the shell is chitinophosphatic in composition. The conularids range from Cambrian to Triassic.

Dentalium

Chiton

Tentaculites

5 cm

Cephalopods

Molluscs with an internal or external shell which may be coiled or straight. The shell is divided into chambers by septa; these are perforated by a hole through which would pass a fleshy tube known as the siphuncle. In the belemnites there is a solid guard. Septal necks (extensions of the septa bordering the siphuncle) may extend backwards from the hole into chambers or camerae. Range Upper Cambrian to Recent.

NAUTILOIDS   A group of cephalopods in which the shell shape is varied. Straight and tightly coiled forms may be discovered but all have straight or gently curved septa. Internal deposits associated with buoyancy are often observed within the chambers and in the tube-like siphuncle formed by the septal necks. Range Cambrian to Recent.

Nautilus   The living *Nautilus* ('pearly nautilus') has a smooth shell in which the body chamber is very large. It covers or overlaps a number of earlier whorls and gives the shell an involute appearance. The septal sutures are gently flexed or folded with a broad lobed appearance. *Nautilus* is first recorded in the Oligocene. It is found as a fossil in Europe, Australia and the East Indies.

Michelinoceras   A long slender shell with a circular section. It tapers slightly towards the rear. Such straight shells are termed orthocones. The siphuncle is placed centrally, and the septal necks are straight. Internal deposits are to be found lining the outer walls and septa. A surface ornament of fine lines may extend around the surface of the shell. *Michelinoceras* extends from the Lower Ordovician to the Upper Triassic. It is recorded from North America, Europe, Asia and Australia.

Orthoceras   *Orthoceras* has a straight, cylindrical shell that expands slightly with growth. The siphuncle is central and remains empty, without secondary deposits. The external surface is smooth or gently furrowed. Septa are simple without folds. *Orthoceras* is known from European and North American Middle Ordovician rocks.

Trochoceras   A slightly irregularly shaped nautiloid that appears to be uncoiling. This type of shell is termed a gyrocone. *Trochoceras* has no siphuncular deposits. Wing-like processes may be present on the shell. *Trochoceras* is inflated and expanded anteriorly. The ornament may consist of fine striae but is usually smooth. The genus *Trochoceras* is confined to the Devonian sediments of Europe.

Dawsonoceras   Cylindrical or subcylindrical in shape, *Dawsonoceras* is characterized by a scalloped or much-lined surface. The siphuncle is also subcentral and the septal necks are short. Some deposits occur within the siphuncle and around the chamber walls. Found in Silurian sediments throughout Europe, Asia, Australia and North America.

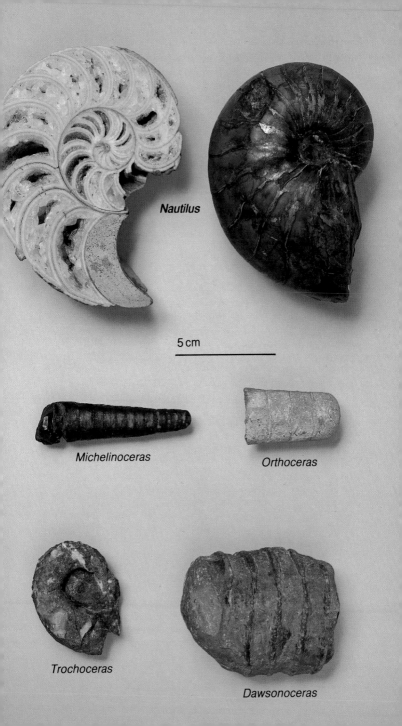

Nautilus

5 cm

Michelinoceras

Orthoceras

Trochoceras

Dawsonoceras

AMMONOIDS The most important of all cephalopods, the ammonoids include goniatites, ceratites and ammonites. They have an external shell that is normally coiled planispirally. The shell is chambered and the suture line folded or fluted into forward-directed **saddles** and backward-directed **lobes**. Internal deposits within the chambers or siphuncle tube are unknown. Range Devonian to Cretaceous.

GONIATITES Goniatite shells are external, small to medium in size and always coiled. The suture lines are comparatively simple, and the folds subdivided into broad, undivided saddles and lobes. Range Devonian to Permian.

Goniatites A small to medium-sized cephalopod with a globular shell. The outer whorl expands anteriorly and overlaps the preceding whorl. The central area or umbilicus is small. *Goniatites* has a rather unique suture line in which the lobes are somewhat angular. It is found only in Carboniferous sediments but is known from Europe, Asia, North Africa and North America.

Gastrioceras This is a member of the same superfamily as *Goniatites*. It has a large umbilicus. The shell is marked externally by strong ribs on the inner margins. The suture line is divided into rounded lobes and saddles, with the ventral lobe narrower than those outside it. *Gastrioceras* is associated with the marine bands of the Upper Carboniferous (Pennsylvanian). It is known from Europe, the USA (Arkansas) and China.

Clymenia This is a member of the important superfamily, the Clymeniaceae whose members are confined to the Upper Devonian. The majority have subdiscoidal shells with a rope-like or evolute coil. *Clymenia* is of small to moderate size with a small umbilicus. The shell has smooth to slightly ribbed ornamentation, and a suture line. *Clymenia* is known from Europe and Australasia.

Beyrichoceras Like *Goniatites* and *Gastrioceras*, this genus is a small to medium-sized cephalopod with subdiscoidal shell. The umbilicus is small. *Beyrichoceras* has a smooth shell, the inner whorls of which taper slightly towards the outer margin. It is found only in Lower Carboniferous (Mississippian) sediments and is known from Europe, North America and North Africa.

5 cm

Goniatites

Gastrioceras

Clymenia

Beyrichoceras

CERATITES   These are coiled ammonoids with small to medium-sized shells. The suture lines are more complex than those of the goniatites, with the lobes subdivided into smaller, secondary folds. Their shells are robust and many species have prominent growth lines and a pronounced ribbing. Range Carboniferous to Triassic.

Ceratites   This genus has a discoidal, although slightly flattened, shell with a wide umbilicus. The outer whorls partly cover preceding ones and there is an increase in whorl height toward the aperture. *Ceratites* has a rather box-like cross-section. The shell has a coarse ornamentation of strong ribs with nodes or tubercules. The ribs do not pass over the venter. Sutures are characteristic; the saddles are rounded and the lobes subdivided or denticulate. *Ceratites* is recorded from Germany, France, Spain and Rumania. It is known only from Middle Triassic sediments.

Joannites   *Joannites* has a smooth, subglobular shell. It is involute, with the outer whorl overlapping or embracing the preceding one. The suture line is more complex than that of *Ceratites*, and is essentially ammonite-like in character, with subdivided lobes and saddles.

Arcestes   *Arcestes* is a small to medium-sized ceratite with an involute, subglobular shell. The suture is ammonitic with triangular-shaped lobes and saddles. Narrow depressed zones or trictions may occur at intervals over the outer surface of the body chamber. The umbilicus is very small or totally enclosed. The shell is usually smooth but fine, closely spaced growth lines may be preserved. *Arcestes* is distributed worldwide but is confined to sediments of Middle and Upper Triassic age. Excellent material exists in Triassic rocks of the European Alps.

Trachyceras   Unlike *Arcestes*, *Trachyceras* has a ribbed and tuberculose shell ornamentation. The shell is involute with a moderate-sized umbilicus and a median furrow on the ventral surface of some species. The sides of the whorls are characterized by slightly flexed ribs which are marked at intervals by small tubercules. The suture is ammonitic but not deeply subdivided as in *Arcestes*. *Trachyceras* is restricted to sediments of Middle and Upper Triassic age. It is known from the European Alps, Spain, Sardinia, Asia, Portuguese Timor and North America. In the USA *Trachyceras* is found in the Middle Triassic of Nevada and California.

Ceratites

Joannites

Arcestes

Trachyceras

5 cm

AMMONITES   Cephalopods in which the suture line is folded into lobes and saddles which themselves are subdivided into minor frills. On the saddles these are termed **folioles**, and on the lobes, **lobules**.

PHYLLOCERATIDS   Many have shells which are tightly coiled with the outer whorl covering the earlier, inner whorls. The suture lines have complex folding with minor frills on both lobes and saddles. Range Lower Triassic to Cretaceous.

Rhacophyllites   A small to medium-sized ammonite, in which the shell is essentially evolute, the outer whorl only partly embracing the inner one. The outer whorl is broader and higher than preceding whorls. Shell smooth with a complex ammonitic suture line. It is recorded from the Upper Triassic sediments of Europe, the Himalayas and the Pacific.

Phylloceras   Involute shell in which the outer whorl covers the preceding one. The shell is compressed and smooth with a small umbilicus, and covered in fine lines. The lobes and saddles of the suture lines are characteristically subdivided with minor frills. *Phylloceras* ranges from the Lower Jurassic into the Lower Cretaceous. It is known worldwide.

Tragophylloceras   A small ammonite with an essentially evolute shell. The outer, ventral half of the last whorl is marked with strong ribs. Large, deep umbilicus. Recorded worldwide from the Lower Jurassic.

AMMONITIDS   This suborder of ammonites is derived from the phylloceratids. It contains many varied species. The suture line is generally less complex than for other ammonites, with some species returning to a 'ceratitic' type suture. Strong keels and ribs are common. Some shells reach gigantic proportions. Range Middle Jurassic to Cretaceous.

Psiloceras   The genus *Psiloceras* is a diagnostic fossil. It is a small to medium-sized ammonite the shell of which is laterally compressed and evolute. Externally the shell is smooth or marked with blunt ribs and closely spaced growth lines. The umbilicus is open. The genus is recorded from the Lower Jurassic of Europe, Asia and North America.

Hildoceras   A medium-sized ammonite with a flattened, evolute shell. The whorls have a rather square cross-section and a ribbed, lateral depression that runs from the aperture towards the umbilicus. A strong keel is present flanked by lateral grooves. *Hildoceras* is restricted to the Lower Jurassic, but is known worldwide.

Dactylioceras   This is possibly one of the best known of all ammonites. It is found worldwide and occurs in great numbers in the Lower Jurassic sediments of the Yorkshire coast, England. The shell is evolute with the numerous whorls marked by a strong ribbing. In some species the ribs are simple but in others they divide over the venter.

5 cm

Rhacophyllites

Phylloceras

Tragophylloceras

Psiloceras

Hildoceras

Dactylioceras

Liparoceras   This is a medium to large-sized ammonite with a robust shell. The whorls increase rapidly in height towards the aperture. The umbilicus is rather wide and deep. A fine to coarse ribbing occurs over the outer surface with two rows of swollen tubercules evident on the side walls. In some species the ribs bifurcate over the venter. *Liparoceras* is restricted to the Lower Jurassic of Europe, North Africa and Indonesia.

Hoplites   *Hoplites* has a rather involute shell which is compressed laterally to give a trapezoidal cross-section. The shell is strongly ornamented with prominent ribs that branch out from a well-developed node. The ribs may be zig-zag in character; they do not cross over the venter. Rib endings tend to alternate with each other in later species. *Hoplites* is an important Middle Cretaceous ammonite, with *H. dentatus* used as a zone fossil for the Lower Middle Albian (Lower Cretaceous) of north-west Europe. *Hoplites* is also known from Mexico and the Middle East.

Douvilleiceras   This genus is characterized by a strongly ribbed shell. The shell is evolute with a rounded cross-section for the outer whorl. The ribs are marked by numerous tubercules. *Douvilleiceras* is recorded from the Albian (Lower Cretaceous) sediments of Europe, North America, Peru, Colombia, Madagascar and India.

Macrocephalites   This genus contains a range of large to gigantic species. The shell is subdivided and evolute, with a wide umbilicus. Inner whorls are slightly compressed. The shells are ribbed, although the outer whorl may become smooth in larger species. *Macrocephalites* is restricted to the lower part of the Upper Jurassic. It is known from Europe, North and South America, Africa and Asia.

Amaltheus   This is a small to medium-sized ammonite with a flattened, discoidal shell. The outer edge has a strong serrated keel and the lateral walls are ornamented with smooth, slightly S-shaped ribs. Some species may have tubercles on the side walls. *Amaltheus* is recorded from the Lower Jurassic of Europe, North Africa and North America. In North America it is found in Northern Alaska, Oregon and Canada.

Perisphinctes   This large to gigantic genus has an evolute shell. The inner whorls are ornamented with many closely spaced ribs, while those of the outer whorl are separated and rather massive. *Perisphinctes* is known from the Upper Jurassic of Europe, Africa, Asia and Cuba.

Liparoceras

Hoplites

Douvilleiceras

5 cm

Macrocephalites

Amaltheus

Perisphinctes

LYTOCERATIDS   It is thought that these 'ammonites' evolved from phylloceratid ancestors in the Upper Triassic. They are again a conservative stock, but lack the leaf-like sutures of their ancestors. The lytoceratid suture is finely branched and the innermost lobe is also subdivided. Many lytoceratid shells have a strong ornamentation and a number exhibit stages of 'uncoiling'. Range Triassic to Cretaceous.

Lytoceras   This genus has an evolute shell in which the whorls enlarge rapidly towards the aperture. The whorls have a rounded or subrounded cross-section. Externally the shell is marked by crinkled growth lines or thin ribs. The suture line is highly complex, with moss-like frills on the lobes and saddles. *Lytoceras* is known worldwide. It has a Lower Jurassic to Upper Cretaceous range. *L. fimbriatum* is a well-known species from the Lower Jurassic of England.

Scaphites   A rather odd-shaped ammonite. The shell is inflated, with the inner whorls coiled and in contact with each other and the other outer whorl hooked at the end of a short shaft. The shell is ribbed and marked with rather elongate tubercles. Scaphites is essentially a mid-Cretaceous form. It is recorded throughout the Northern Hemisphere and from Madagascar and Australia.

Hoploscaphites   This genus is clearly related to *Scaphites* and may be termed heteromorphic. The initial whorls overlap but the last has become hook-like at the end of a very short shaft. The shell is rather flat-sided with distinct ridges and tubercles. *Hoploscaphites* is restricted to the Upper Cretaceous sediments of Europe, North America, South America (Chile) and Antarctica.

Turrilites   This resembles a large to very large gastropod. It is, however, a lytoceratid with a distinctly heteromorphic shell. It was, like many other heteromorphs, a sea floor dweller. The shell has an acute apical angle and the whorls are tightly coiled in a helical manner. Ribs and tubercles ornament the shell. *Turrilites* is recorded from the Upper Cretaceous sediments of Europe, Africa, India and Japan. In the USA it is known, along with other members of the Turrilitaceae, from the Cretaceous of Texas.

Hamites   *Hamites* and *Scaphites* are both examples of heteromorphic ammonites. The shell of *Hamites* is coiled in an open plane spiral so that chambers occur along three separated subparallel shafts. Like *Scaphites* it is distinctive from other, more normal, ammonites. *Hamites* has a ribbed ornamentation. The ribs are usually closely packed but may be separated or distant. They circle the whorls in all species. *Hamites* is known from the Lower Cretaceous of the Northern Hemisphere. It is particularly common in the Gault sediments of England.

5 cm

Lytoceras

Scaphites

Hoploscaphites

Turrilites

Hamites

BELEMNITES   These fossil cephalopods are grouped with the living squids and octopuses. The internal skeleton is divided into a long, heavy bullet-like **guard**, a chambered, conical **phragmacone** and a tongue-like **pro-ostracum**. Of these, the guard is the most resistant part and numerous specimens may be found in Jurassic and Cretaceous rocks. Range Carboniferous to Tertiary. Although distributed worldwide the belemnites were more common in the northern 'Boreal' realm than in the southern 'Tethyan' area.

Hibolites   This genus has an elongate guard with a bulbous posterior region. The guard tapers anteriorly and a distinct groove occurs on the ventral surface. *Hibolites* has a marked spear or club shape. It ranges from the Jurassic into the Lower Cretaceous.

Actinocamax   This genus has a cylindrical guard with a deep ventral furrow. The guard is rounded to subrounded in section, with a distinct funnel-shaped notch anteriorly for the placement of the phragmacone. The notch is relatively shallow compared with other genera. A small pointed extension occurs at the posterior end of the guard. *Actinocamax* is recorded worldwide from sediments of the Middle and Upper Cretaceous.

Neohibolites   *Neohibolites* has a small guard with a circular cross-section. A short slit and groove are present near the anterior notch. Both occur on the ventral surface. *Neohibolites* is known only from the Upper Cretaceous of Europe.

Belemnitella   This has a medium to large-sized guard which is frequently marked with lateral lines. The sides of the guard may have a granular appearance. It is subcylindrical in shape, slightly flared anteriorly, and has a blunt tip. The anterior notch is deep and steeply inclined towards the ventral surface. Good specimens may exhibit traces of blood vessels over the outer surface. A long slit may occur on the ventral surface. *Belemnitella* is restricted to the Upper Cretaceous of Europe.

Cylindroteuthis   This genus has a large cylindrical guard, with an oval cross-section. The sides may be slightly flattened and the ventral surface is noted for the presence of a strong groove. The latter tends to deepen towards the tip of the guard. *Cylindroteuthis* ranges from Early Jurassic to the end of the Cretaceous. It is known only from Europe and North America.

Hibolites

Neohibolites

Actinocamax

Belemnitella

belemnite section
showing conical
phragmacone
and guard

5 cm

Cylindroteuthis

Arthropods

The largest of all animal groups. It includes insects, spiders, scorpions, crabs, lobsters, shrimps, prawns, barnacles, ostracodes, millipedes, centipedes and several important groups of fossils. The arthropods have a hard outer skeleton (exoskeleton) and jointed limbs. Most arthropod groups are poorly represented in the fossil record, but the trilobites of the Palaeozoic are the important exception to this rule. Range Cambrian to Recent.

Crustaceans

Although fossil crustaceans are rare in comparison with trilobites, a number of genera are common in Cretaceous and Cainozoic formations. Range Cambrian to Recent.

Hoploparia A small lobster with an elongate but slightly depressed body. The skeleton is divided into three distinct regions: head, thorax and abdomen. The rostrum over the head is long and narrow. Long legs and large pincers are typical of *Hoploparia* and related genera. The genus is recorded worldwide from Cretaceous and Cainozoic sediments.

Beyrichia This has a two-valved skeleton that would normally enclose the soft parts. The ostracodes are microscopic in size, with the majority of individuals measuring between 0.5 and 5 mm ($\frac{1}{50}$ and $\frac{1}{5}$ in). The valves are of slightly unequal size and articulate along a straight hinge line. Hinge teeth are present in one valve and sockets in the other. *Beyrichia* is a strongly ornamented genus with a nodular or granular surface texture. *Beyrichia* is recorded from the Lower Silurian to Middle Devonian sediments of Europe, North America and Australia.

Balanus The barnacles or cirripedes are highly specialized crustaceans. The skeleton is modified into a protective 'shell' composed of several large plates. The 'shell' is usually cone-shaped, with a lid. The barnacles range worldwide from the Ordovician to the present day.

Notopocorystes This Cretaceous crab has an ovoid to shield-shaped carapace which is usually longer than it is broad. The surface is often finely granulated with a median ridge of small nodules or tubercles. Known from Europe, North America, and the Middle and Far East.

Aeger This crustacean is superficially similar to the living shrimps and prawns. *Aeger* is, however, a member of an extinct group of swimming decapods. It has a compressed body and a long projecting beak in front of the eyes. *Aeger* is common in the Triassic and Jurassic.

Insects

Insects appear first in Silurian/Devonian deposits. Rare as fossils.

Insect in amber Amber is the fossilized resin of pines and other gymnospermous trees. Insects are often trapped on the sticky surface of the tree and gradually become encased in amber.

Hoploparia

Beyrichia
(×15)

Balanus

Notopocorystes

Aeger
(×0.1)

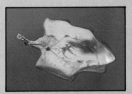

Insect in amber
(×0.5)

5 cm

Trilobites

The body of these extinct arthropods is divided into distinct head (**cephalon**), thorax and tail (**pygidium**) regions. Two furrows extend along the length of the body separating a central axial area from two lateral areas. The head is usually crescentic or semicircular, with a pronounced central region called the **glabella**. Lateral cheeks bear the eyes. Facial sutures divide the cheeks into free (outer) or fixed (inner) cheek areas. **Genal spines** may extend backwards from the cephalon. The thorax and pygidium are segmented. Range Cambrian to Permian.

Olenellus This trilobite has a broad, semicircular–shaped cephalon with short genal spines. The glabella has short furrows present on each side. *Olenellus* has 14 spiny thoracic segments and a minute pygidium which is overlapped by a long axial spine. *Olenellus* is recorded from the Lower Cambrian sediments of Scotland, Greenland and North America.

Peronopsis This genus is a diminutive trilobite with headshield and tail of equal size. The cephalon has a distinct rim around the edge, and the glabella is squared off in outline. *Peronopsis* has two thoracic segments. The pygidium has a broad, rather triangular–shaped axis and a distinct rim. *Peronopsis* is restricted to sediments of Middle Cambrian age and is found in North America (Montana), Europe and China.

Trinucleus This genus has a relatively large cephalon that is broader than long. The glabella expands towards the front of the cephalon. A fringe of well-developed pits occur around the front and sides of the headshield. Long genal spines, approximately twice the length of the thorax and pygidium, are evident. The thorax has six segments. Lower and Middle Ordovician of the British Isles and Scandinavia.

Ampyx A small trilobite with a long spine that projects forward from the inflated glabella. Only six thoracic segments are present. The pygidium is short and broad. *Ampyx* is known only from the Middle Ordovician of North America and Europe.

Encrinurus This genus has a large, inflated glabella covered in small pustules. These also cover the cheek areas and give the cephalon a rough texture. The eyes are small and stalked. There are 11 or 12 thoracic segments and a sharply tapered pygidium. The pygidium is clearly segmented and has an ornament of three rows of small tubercles. *Encrinurus* is a cosmopolitan genus from Middle Ordovician to Silurian.

Dolichoharpes This unusual trilobite has a large semicircular to oval cephalon. The glabella is strongly convex with well-developed furrows. Large, long genal spines extend the whole length of the body and tail. The border of the cephalon is ornamented with large pits. Numerous body segments occur. The pygidium is short and wide. *Dolichoharpes* is recorded from the Middle Ordovician of North America and Ireland.

Olenellus

Peronopsis

Trinucleus

Ampyx

Dolichoharpes

Encrinurus

5 cm

Ellipsocephalus   *Ellipsocephalus* has a suboval-shaped cephalon with rounded genal angles. The glabella is long, with up to three pairs of lateral furrows. There are 12 to 14 thoracic segments and these are rounded off laterally. The pygidium is small and fused. *Ellipsocephalus* has a facial suture that cuts the rear margin of the cephalon inside the genal angle; this type of suture is termed opisthoparian. *Ellipsocephalus* is recorded from Lower and Middle Cambrian sediments. It is known from Europe, Morocco and possibly Australia.

Angelina   *Angelina* is a member of the same superfamily as *Olenus* (*see* below). It has an opisthoparian facial suture. *Angelina* is much larger than *Olenus*, however, with a narrow glabella and long genal spines. The thorax is comprised of numerous segments, the borders of which are rather spinose. *Angelina* is restricted in time to the Lower Ordovician and is known only from Wales and possibly eastern Canada.

Olenus   *Olenus* is a small trilobite with a diminutive tail. The cephalon is broad and rather narrow. A subcylindrical glabella is marked by two or three pairs of lateral furrows. The suture line cuts the posterior border of the cephalon inside the short genal spines. There are 13 to 15 thoracic segments. *Olenus* is recorded from the Upper Cambrian of northern

Ogygiocaris   The pygidium of this genus is approximately the same size as the cephalon. The latter is semicircular with short genal spines. The eyes are crescent-shaped and the suture opisthoparian. There are eight thoracic segments. *Ogygiocaris* is locally abundant in the Lower and Middle Ordovician sediments of northern Europe.

Bumastus   *Bumastus* is a rather unusual trilobite. It has a compact skeleton with cephalon and pygidium of subequal size. Both are smooth and rounded and unsegmented. The suture line is short, and starts on the side of the cephalon just above the level of the eye. There are ten thoracic segments. *Bumastus* is a cosmopolitan species from Upper Ordovician and Silurian sediments.

Pseudogygites   This is a medium-sized trilobite with an opisthoparian facial suture. The cephalon and pygidium are of subequal size and there are eight thoracic segments. The cephalon is shovel-shaped with small crescentic eyes and short genal spines. A broad axial region is characteristic of this genus. The outer margin of the pygidium is depressed and marked with thin concentric lines. *Pseudogygites* is known from the Middle and Upper Ordovician of Canada. Excellent specimens occur near Bowmanville, Ontario.

5 cm

Angelina

Ellipsocephalus

Olenus

Ogygiocaris

Bumastus

Pseudogygites

Calymene   This well-known trilobite is medium-sized, with the head much larger than the tail. The head is semicircular with a strongly convex glabella which tapers anteriorly. It is also lobed, with the largest lobes occurring toward the rear. The facial sutures pass from the front edge of the cephalon around the back of the small eyes to bisect the genal angle. Normally there are 13 segments in the thorax. The pygidium is fused although there are six distinct rings on the axial region. *Calymene* ranges from Lower Silurian to Middle Devonian. It is cosmopolitan in distribution.

Flexicalymene   This genus, and the related *Calymene*, belong to the same family of trilobites. Small differences in structure separate the two but *Flexicalymene* is often found enrolled with the tail tucked beneath the headshield. This is for protection, the enrolled animal being less vulnerable to attack. *Flexicalymene* has three pairs of lobes on its glabella, with the back pairs having a circular appearance. The pygidium is deeply grooved. *Flexicalymene* predates *Calymene*. Excellent material has been recovered from the Middle and Upper Ordovician rocks of Europe and North America (e.g. Spain, and Ohio, USA).

Phacops   This genus has a compact skeleton with a large head and moderately sized tail. The cephalon is dominated by the glabella, which is inflated and strongly convex. *Phacops* has very large eyes and proparian facial sutures, and it has 11 thoracic sutures and a fused pygidium. Many specimens are found rolled into a tight ball. *Phacops* is a cosmopolitan trilobite with a Silurian to Devonian range.

Dalmanites   *Dalmanites* is related to *Phacops*. It is usually larger, however, and has several distinguishing features. The cephalon is semicircular with long genal spines. A short spine is also present at the front of the cephalon. The glabella expands anteriorly and the eyes are large and crescent-shaped. *Dalmanites* has 11 thoracic segments. The pygidium is large with a long spine. *Dalmanites* is known from Europe, North America and South America. It has a Silurian to Lower Devonian range.

Greenops   The headshield of *Greenops* is similar to that of *Phacops*, in that it has an inflated glabella and large crescentic eyes. *Greenops*, however, has large genal spines and a unique pygidium; this is relatively small with a much segmented axial region. The border is distinctly toothed in appearance. *Greenops* is recorded from the Middle Devonian sediments of Europe and North America.

Calymene

Flexicalymene

Phacops

Dalmanites

Greenops

5 cm

Echinoderms

These invertebrates have skeletons made up of calcareous plates formed beneath the outer layer of soft tissue. Range Cambrian to Recent.

Crinoids

The sea-lilies have a cup-shaped body (the **calyx**) consisting of several regularly arranged rings of calcareous plates. Most crinoids also have segmented arms and stems composed of disc-like **ossicles**. Range Cambrian to Recent.

Amphorocrinus   This genus has a low, ovoid to spherical calyx which consists of three basal plates and a second row of six plates; two rows of arm plates occur above. A raised area or tube-like extension composed of small hexagonal plates covers the upper surface. Ten to 30 biserial arm branches may be found associated with the calyx. *Amphorocrinus* is known from the Upper Carboniferous of Europe and North America.

Phanocrinus   This rather small crinoid from the Carboniferous of North America has a calyx of three rows of large, fused plates. The five pairs of thick, strong arms consist of columns of rounded ossicles.

Stellarocrinus   This is a small, compact crinoid. It has a bowl-shaped calyx, consisting of three rows of plates. The stem is quite robust and the arms consist of two rows of plates (biserial). They branch just above the calyx. *Stellarocrinus* is recorded from the Upper Carboniferous (Pennsylvanian). It is known only from North America.

Abatocrinus   This genus has a small, rounded or top-shaped calyx comprised of three rows of plates, of which the basals are the largest. An upper covering is composed of hexagonal plates. The arms appear to arise from small holes at the boundary between the cup and the covered tegmen. Recorded worldwide during the Lower Carboniferous.

Marsupites   This is a large, stemless crinoid. It has a large plated calyx comprised of three rows of five plates. The base of the stem is occupied by a single large plate. The plates have a pattern of well-defined ridges. The arms of *Marsupites* are short and rather small. *Marsupites* is known from the Upper Cretaceous of Europe and North America.

Apiocrinus (Apiocrinites)   This crinoid has a large, streamlined calyx, the upper ossicles of the stem and the basal plates being fused together. The radial plates are relatively small, and above them four rows of brachials occur. These are also fused to the calyx. Ten arms are present in perfectly preserved specimens. *Apiocrinus* is restricted to Europe and is known only from Jurassic sediments. Crinoid holdfasts are quite rare; the example shown is in the form of an external cast.

Phanocrinus

Stellarocrinus

Amphorocrinus

Marsupites

Abatocrinus

Apiocrinus

Holdfast

5 cm

Echinoids

The echinoids have a strong test consisting of numerous plates. These are arranged into ten radial 'segments' that extend around or over the test. The mouth is found on the lower (**adoral**) surface, while the anus is often an integral part of a complex circular arrangement of plates called the **apical disc**. This is found on the upper (**aboral**) surface. The ten 'segments' are separated into the narrower, pore-bearing **ambulacra** and the broader **interambulacra**. Both mouth and anus may have migrated from their original positions in certain species. The echinoid test may be globular, conical or disc-shaped. Range Ordovician to Recent.

Cidaris   This may be defined as a regular echinoid – one in which the mouth and anus are in the central position on the lower and upper surfaces respectively. Cidaroid echinoids range from the Upper Silurian to the present day. *Cidaris* has a robust, heavily ornamented test. It is subrounded in shape with a large tubercle for the location of a strong spine located on each of the interambulacral plates.

Phymosoma   This is a small to medium-sized regular echinoid with a flattened, discoidal test. It has a strong external ornamentation of closely spaced tubercles. The plates surrounding both mouth and anus are invariably missing and give the false appearance of large ventral and dorsal openings. Small ridges or striae radiate outwards from the central raised area of the larger tubercles. *Phymosoma* is recorded from the Upper Jurassic sediments of Europe, North Africa, India and the Americas.

Hemicidaris   Like *Phymosoma*, this genus has a small to medium-sized test. It is, however, slightly higher, although the underface is flattened. The interambulacral tubercles are strongly developed, and are known to bear elongate spines. The ambulacral areas also have raised tubercles, particularly towards the ventral portion of the test. *Hemicidaris* ranges from the upper Jurassic into the Upper Cretaceous and is cosmopolitan in distribution.

Pygaster   This medium-sized irregular echinoid has a depressed, somewhat flattened test. Unlike *Cidaris* the anus of *Pygaster* has moved slightly away from the central position. *Pygaster* has only small tubercles that support short weak spines. The five ambulacra are straight and radiate outwards from the apical area. *Pygaster* ranges from the Upper Jurassic into the Upper Cretaceous; it is restricted geographically to Europe.

Acrosalenia   *Acrosalenia* is a small echinoid with a flattened test. It has large tubercles on the plates of the interambulacral areas. The anus is slightly displaced from the centre of the apical area. Two rows of small tubercles occur on each of the ambulacral areas; these are rather sinuous in appearance. *Acrosalenia* is known from the Jurassic and Cretaceous sediments of Europe and East Africa.

5 cm

Cidaris

Phymosoma

Hemicidaris

Pygaster

Acrosalenia

Clypeaster   This echinoid has a medium to large test which is flattened with a rounded margin. The mouth is almost central on the ventral surface but the anus has migrated with a long groove marking its presence on the upper surface of the test. *Clypeaster* rests partially buried on the sea floor, and the ambulacral areas extend only to the margin of the test. The interambulacral areas are very wide and lack the robust tubercles of other genera. *Clypeaster* ranges worldwide from the Upper Jurassic to the present day.

Holectypus   *Holectypus* is a medium-sized irregular echinoid. The test is flattened on the undersurface but overall is hemispherical or globular in shape. Both mouth and anus occur on the ventral surface. The ambulacral areas radiate outwards from the apical area and may be traced around the test to the mouth. No strong tubercles are evident on either ambulacral or interambulacral plates. *Holectypus* ranges throughout the Jurassic and Cretaceous of both Europe and North America.

Conulus   This small, irregular echinoid has a highly conical test. The undersurface is flattened and supports both mouth and anus. Numerous small tubercles occur on the interambulacral areas. The ambulacra are rather narrow and straight. *Conulus* is known from the Upper Cretaceous of Europe, North America, North Africa and Asia.

Micraster   *Micraster* is one of the best-known of fossil echinoids. It has a characteristic heart-shaped test with a flattened undersurface. The test has a slightly inflated appearance with the posterior region of the test higher than the front. Five petaloid (petal-shaped) arms occur on the upper surface; they radiate from the centre of this surface and are indented. The anterior ambulacrum runs forward into a deep groove. The mouth has moved towards the front of the anterior surface, and the anus is sited posteriorly. *Micraster* is used as a zone fossil in determining the Upper Cretaceous rocks of Europe. It is recorded from Europe, the Mediterranean Sea, Cuba and Madagascar.

Nucleolites   *Nucleolites* is related to *Clypeaster*. It is much smaller, however, and the test is slightly more inflated. The test is subcircular in outline with a prominent notch posteriorly; this contains the anus. The mouth has moved towards the front of the undersurface of the test. The five ambulacral areas are slightly petaloid in appearance. *Nucleolites* ranges from the Middle Jurassic to the Upper Cretaceous. It is known from Europe and Africa.

Scutella   A medium to large-sized irregular echinoid with a flattened test which is braced internally by supports or buttresses. The ambulacra are short and strongly petaloid in shape. On the undersurface the mouth is central and the anus placed marginally. Feeding grooves extend from the edges of the test in towards the mouth. The test is semicircular in shape. Found in Miocene sediments of Europe and north Africa.

5 cm

Clypeaster

Holectypus

Micraster

Conulus

Nucleolites

Scutella

Blastoids

The blastoids are echinoderms comprised of a bud-like calyx and a well-developed stem or stalk. They show a pronounced radial symmetry with five elongate petal-shaped ambulacra. The calyx or **theca** consists of 13 plates. The blastoids were exclusively marine in habitat and are mostly found in shallow-water limestones. Range Silurian to Permian.

Pentremites *Pentremites* has a clearly defined radial symmetry. It has five ambulacral rays and five spiracles or outlets around the mouth. The calyx is bud-like and comprises a limited number of large plates. Each ambulacrum forms a V-shaped depression or groove. The calyx is the more resistant part of the skeleton. *Pentremites* is restricted to North and South America, and is known only from Carboniferous sediments.

Orbitremites A small, rounded blastoid with narrow, slightly flared ambulacra that stand out from the body of the calyx. The interambulacra are broad and consist of three large plates. Five spiracles are present around the mouth. One of these is enlarged to include the anus. *Orbitremites* is known from Lower Carboniferous (Mississippian) to Permian sediments of Europe and North America. Excellent specimens are recorded from the Osagian (Lower Mississippian) carbonate sediments of Iowa, USA.

Cryptoblastus A small, melon-shaped blastoid with five tiny spiracles on the upper surface. The ambulacra are less distinct than those of *Pentremites* but the blastoid radial symmetry is still clearly defined. *Cryptoblastus* is known from the Lower Carboniferous of North America.

Cystoids

Members of this group of echinoderms have sac-like or spherical thecae in which the number of individual plates may exceed 1500. These are irregularly arranged and often pierced by a variety of pores. Arms may or may not be present. The cystoid stem varies in length and it was probably used in some cases to help the animal move over the sea floor.

Holocystites *Holocystites* has a rather elongate, sac-like theca. It is composed of several regular rows of plates. These are relatively large and invariably six-sided. The plates at the upper end of the theca are smaller than those of the main body. *Holocystites* is a member of the Rhombiferida, a group of cystoids characterized by the rhomb or diamond-shaped outline of the pore-bearing areas. Some 50 species of *Holocystites* are recorded from the Silurian sediments of central USA.

Globoblastus A small to medium-sized blastoid with a sub-rounded theca; a prominent sieve-like plate is present on the upper surface. It is only common from the Lower Carboniferous of North America.

5 cm

Pentremites

Orbitremites

Cryptoblastus

Holocystites

Globoblastus

Asterozoans/Ophiuroids

These are better known as the starfish and brittlestars. They are comparatively rare as fossils although they first appeared during the Ordovician. Occasionally specimens are found as casts and impressions in clays and other fine sediments. Range Ordovician to Recent.

Metopaster This is a moderate to large-sized starfish. It lacks distinct arms, and has a pentagonal shape. The main body consists of numerous tiny plates. On the upper surface these are rather random in their organization, whereas those on the undersurface are more regular and distinct ambulacral and interambulacral areas can be distinguished. *Metopaster* is characterized by an outer rim of large marginal plates. These are often found isolated in the Cretaceous sediments of northern Europe.

Palaeocoma *Palaeocoma* is a typical representative of the brittlestars. It has a well-defined central disc and five long, thin arms. The mouth is present on the undersurface. Ophiuroids do not have an anus. The mouth is flanked by five distinct buccal plates. On the undersurface, the inner or proximal areas of the arms extend inwards to reach the mouth. *Palaeocoma* and related genera are known from Jurassic to Recent sediments. The living *Ophioderma* is descended from this genus. Brittlestars are known worldwide.

Lapworthura This brittlestar has a large central disc and robust arms. The mouth is placed centrally within a star-shaped arrangement of small plates. As with *Palaeocoma* the arms extend over the undersurface of the disc to make contact with the oral region. The arms are broad, with long vertical spines. *Lapworthura* is found in the Upper Ordovician and Silurian sediments of the British Isles and Australia. Good material is recorded from the Middle Silurian of the Welsh borders of Great Britain.

CARPOIDS The carpoids are small to moderately sized echinoderms. They have a well-defined theca with large plates on the upper surface. These plates are usually arranged along the long axis of the body, but the well-defined radial symmetry of other echinoderms is lacking. Some carpoids have a well-developed stem or 'tail' and short arm-like structures. The majority are free-living creatures that rested or crawled over the sea floor. It is thought that the carpoids, or calcichordates as they are often called, were ancestors of the fishes. Range Middle Cambrian to Lower Devonian.

Lagynocystis This is a small calcichordate with a flattened, sac-like body. It is longer than it is wide and very asymmetrical. A single brachiole occurs on the front left-hand corner of the theca. The stem or tail is much reduced and may be totally absent. The pyramid-shaped species *L. pyramidalis* is known from the Middle Ordovician of Bohemia. Doubtful lagynocystids are recorded from the Welsh borders of Great Britain.

Metopaster

Palaeocoma

5 cm

Lapworthura

Lagynocystis

Graptolites

Graptolites are colonial marine animals that existed during the Palaeozoic era. The colonies (**rhabdosome**) consist of one or more branches (**stipes**), each of which supports a number of cup-like structures (**thecae**). Two types of thecae are present in the shrub-like dendroid graptolites, and one in the graptoloids. The graptolites range from Upper Cambrian to Upper Carboniferous.

Dendroid graptolites

Dictyonema   This is one of the best-known of dendroid graptolites. It has a conical-shaped rhabdosome consisting of numerous delicate stipes. Numerous tiny thecae occur on each stipe. The genus is recorded worldwide from deep-water sediments of Upper Cambrian to Carboniferous age.

Graptoloids

Phyllograptus   *Phyllograptus* is a member of the tetragraptid family. The colony consists of four branches but in this genus they are united along their length. In cross-section the colony is therefore cross shaped. The thecae are long and slightly curved, and arranged in a 'back-to-back' formation. The vast majority are flattened during the process of preservation. *Phyllograptus* is therefore leaf-like in appearance. The genus is locally abundant in the black shales of the Lower Ordovician. It is known worldwide.

Tetragraptus   This graptolite has a rhabdosome made of four stipes which branch in pairs. The colony has a bilateral symmetry. Each stipe is characterized by the presence of numerous closely packed, tooth-like thecae. This tends to give the stipe a saw-like appearance. Four branches are diagnostic. *Tetragraptus* is recorded from Lower Ordovician black shales, throughout the world.

Dicranograptus   The genus *Dicranograptus* is noted for the division of its colony into two distinct parts. The lower area consists of a single branch lined on either side by a row of thecae. Above, the colony divides into two branches each with a single row of thecae cups. The latter are slightly curved or rounded in shape. *Dicranograptus* ranges from Lower Ordovician into Upper Ordovician and is recorded from Europe, North America, South America, Australia and Asia.

Didymograptus   The various species referred to this genus have two stipes. These may hang down like the two prongs of a tuning fork, or spread outwards to form an almost straight line. The thecae are usually simple, tooth-like or slightly curved. Two important 'tuning-fork' species occur in Lower Ordovician rocks. The most common of these is *D. murchisoni*, which is much larger than the related species *D. bifidus*. The genus is known worldwide during the Lower Ordovician.

5 cm

Dictyonema

Tetragraptus

Phyllograptus

Dicranograptus

Didymograptus

Orthograptus This genus has a single branched colony in which the thecae are arranged on either side of the stipe (**biserial**). The thecae are comparatively large and either straight or slightly curved. Small spines occur on the upper outer edges of the cups. *Orthograptus* is known from Upper Ordovician to Lower Silurian sediments and it is distributed worldwide.

Climacograptus Like *Orthograptus* this is a biserial graptoloid. The thecae are large and slightly incurved towards the stipe. They are also offset on either side. The colony is elongate, circular in cross-section, and noted for the presence of a double spine structure at the beginning of the colony. *Climacograptus* is known worldwide from rocks of Ordovician and Lower Silurian age. Excellent material has been collected from the Birkhill Shales (Lower Silurian) of Scotland and the Normanskill Group (Middle Ordovician), New York State, USA.

MONOGRAPTIDS The monograptids are characterized by a single branched colony with stipes arranged along one edge only (uniserial). Individual species are identified on the shape and size of the thecal cups, which range from having a simple toothed appearance to being incurved or lobed. The monograptids are diagnostic of the Silurian in North America and the British Isles. They are abundant in areas as far apart as the Baltic region of Sweden, southern Scotland, and Oklahoma, USA.

Rastrites *Rastrites* is a monograptid. It has a single stipe and the thecae are arranged in a uniserial manner. The stipe is slightly curved, however, and the thecae are long and tube-like. They are also distanced from each other and give the colony an 'open' appearance. The initial part of the colony may be tightly coiled. *Rastrites* is confined to sediments of Lower Silurian age. It is known worldwide with the exception of South America (and possibly North America).

Cyrtograptus This genus has a similar geographical distribution to *Rastrites*, except that good examples are known from the Middle Silurian (Niagaran) of Idaho, USA. *Cyrtograptus* is related to *Monograptus* but has several branches. The main stipe is coiled and numerous subsidiary branches radiate outwards. Numerous thecae occur along the outer edge of each stipe. *Cyrtograptus* is diagnostic of the Middle Silurian.

Climacograptus

Orthograptus

Rastrites

5 cm

A. Cyrtograptus
B. Orthograptus Nilssoni
Wenlockian
Birnieth shale, Penang

A

B

Cyrtograptus

Vertebrates

The vertebrates have internal skeletons of cartilage or bone. Some, like early fish, turtles and armadillos, have a heavy protective armour which may either cover the head or encase the body. The most common finds are individual teeth and bones. Range Lower Ordovician to Recent.

Fish

The earliest fossil fish appear during the Lower Ordovician. Bony fish and fish with cartilaginous skeletons appeared in Silurian to Devonian times.

ARMOURED FISH

Cephalaspis The head of this genus is covered by a heavy bony shield, and the body by thick scales. Prominent 'horns' or spines extend backwards from the corners of the headshield. The eyes occur in the centre of the dorsal surface. *Cephalaspis* is known from the Upper Silurian and Devonian of Europe and North America.

SHARKS AND RAYS

Ptychodus This shark has flattened teeth that are used for crushing shell fish. The teeth are rather square in shape with a strongly ridged upper surface. *Ptychodus* is recorded from the Cretaceous sediments of Europe, North America, Africa and Asia.

Odontaspis *Odontaspis* is a living sand shark. It is first recorded from the Upper Cretaceous of Europe but is known worldwide in later sediments. Skeletal remains are confined to the teeth. These are high and sharply pointed with a small subsidiary point on either side.

Myliobatis The eagle rays, like the sharks, have a cartilaginous skeleton. Their teeth are common in Upper Cretaceous to Tertiary sediments. The teeth are wide and flattened with several teeth joined to form a tooth plate. *Myliobatis* is a cosmopolitan genus.

BONY FISH

Pycnodus As with other vertebrates that crush their food, the teeth of *Pycnodus* are flattened and robust. They form plates or pebbly pavements on the inside of the mouth. *Pycnodus* is a deep-bodied fish, many species of which occupied niches in reefs and rocky coastal environments. It ranges from Jurassic into Eocene and is distributed worldwide.

Lepidotes This genus has a short mouth, lined by strong teeth. The tail is symmetrical and the body scales thick and shiny with a characteristic diamond shape. The scales are the most common fossils associated with this bony fish. *Lepidotes* is known from the Mesozoic of Europe, Africa and North America.

5 cm

Cephalaspis
headshield

Ptychodus
palate teeth

Odontaspis
teeth

Myliobatis
ray teeth

Pycnodus
palate teeth

Lepidotes scales

Reptiles and birds

In the stratigraphic record the major groups of fossil reptiles are turtles, crocodiles, ichthyosaurs, plesiosaurs and dinosaurs.

ICHTHYOSAURS   The ichthyosaurs were the dominant marine reptiles of the Mesozoic Era. They are known through isolated bones or, more rarely, complete skeletons. Vertebrae are perhaps the most common finds. They are usually biconcave with two closely spaced projections on the upper surface. Ichthyosaur teeth are long and pointed with deep vertical grooves. Ichthyosaur remains are known worldwide.

PLESIOSAURS   Isolated bones and complete skeletons of plesiosaurs are locally abundant in Jurassic and Cretaceous clays and shales. Plesiosaur vertebrae are usually larger and flatter than those of ichthyosaurs. The teeth are long and pointed but most lack the vertical grooves present in ichthyosaur teeth. The plesiosaurs are recorded worldwide.

TURTLES   Turtles first appear in rocks of Triassic age. They are locally abundant in Jurassic, Cretaceous and Tertiary strata.

Puppigerus   In contrast to *Trionyx*, *Puppigerus* is a marine turtle with a smooth shell. The shell is divided into numerous plates that form a characteristic mosaic. *Puppigerus* is restricted to the Eocene of Europe and North America.

Trionyx   The strongly ridged or pitted scutes of *Trionyx* are abundant in sands and clays of Tertiary age. Trionychids first appear in the Cretaceous and are recorded from Europe, North America, Africa and Asia.

CROCODILES   Crocodiles are among the commonest fossil reptiles. They first appear in rocks of Triassic age, and different families are locally abundant in Jurassic, Cretaceous and Cainozoic strata. The most common finds are of heavily pitted scutes, stout vertebrae, and teeth with sharply pointed crowns.

DINOSAURS   The remains of dinosaurs are common in some parts of North America, Europe and Africa. They are recorded worldwide during the Mesozoic, but good finds are rare. *Iguanodon* from the Lower Cretaceous is known through complete skeletons, but its flattened teeth and vertebrae are more likely finds.

BIRDS   Birds originated in the Jurassic. Their bones are light in weight and hollow and thus extremely delicate; their fossil record is very poor. The articulations between limb bones have a characteristic appearance.

vertebra

Ichthyosaur
vertebrae

tooth

Plesiosaur

Trionyx
carapace bone

Puppigerus
carapace bones

tooth

Crocodile

rib

tooth
fragment

Iguanodon

Bird
foot

5 cm

Mammals

Although the mammals first appeared during the Triassic they are poorly known until the Cainozoic. Most mammals are identified by the form of their cheek teeth. Plant-eating mammals (**herbivores**) have high-crowned cheek teeth. These are usually square or rectangular with flattened cusps. Meat-eaters or **carnivores** have low, narrow crowns in which the cusps are pointed and sharp.

PLANT-EATING MAMMALS

Equus   The generic name *Equus* embraces the horse, donkey and zebra. The teeth are very tall, with deep lateral grooves. Upper cheek teeth have square crowns, whereas those of the lower jaw are rectangular. Both upper and lower teeth have a complex pattern of short crescentic crests. The first horses appear in the Eocene. They are known from Europe, North America, Africa and Asia.

Caenopus   *Caenopus* was a hornless rhinoceros of fairly large size. The skull was approximately 30 cm (12 in) long and the cheek teeth had developed with crescentic crests. It is restricted to North America and is only known from Lower and Upper Oligocene sediments.

Elephas   The elephants have very large cheek teeth which in side view appear deeply grooved. The upper surface is flat with subparallel ridges across the width of the tooth. Elephants are first recorded from Pleistocene sediments. They are known from Europe, Asia and Africa. The name *Elephas* applies only to the Indian elephant. *Loxodonta* is relatively common in the Oligocene strata of South Dakota, USA.

Merycoidodon   *Merycoidodon* was a pig-like mammal with a short, relatively deep skull. The tooth crowns are quite low and square in outline. Crescentic cusps are present on their upper surfaces. *Merycoidodon* is restricted to the Oligocene of North America and is locally abundant in South Dakota.

MEAT-EATING MAMMALS   Carnivorous mammals really expanded during the Early Cainozoic, with the cats, dogs and bears appearing throughout the Tertiary.

Canis   The generic name *Canis* embraces the domestic dog, the wolf and the dingo. Dogs have sharp, pointed canine teeth and their cheek teeth have a sharp cutting edge that is used to slice through flesh and crush bones. The dogs are known worldwide. They first appeared in the Pliocene.

RODENTS   The rodents are small mammals whose jaws and teeth are specialized for gnawing and nibbling their food. The rodents arose from the squirrel-like *Paramys* that lived during the Palaeocene.

Ischyromys   This genus is an early squirrel. It has moderately high-crowned teeth that have proud, crescent-shaped cusps. In side view the teeth flare slightly towards the chewing surface, and are ridged. *Ischyromys* is common in the Oligocene of South Dakota, USA.

Equus tooth

Caenopus tooth

Elephas teeth

Merycoidodon

Ischyromys jaw fragment with teeth

Canis

5 cm

Plants

Although fossil plants are not as common as fossil animals, their remains are locally abundant throughout the stratigraphic record. The earliest evidence of primitive plants occurs in the Precambrian, with algal stromatolites among the earliest of fossils. Plant spores and pollen are also important in the dating and correlation of some rocks. The plants may be divided into vascular and non-vascular varieties, and the vascular plants (those with tube-like tissues that carry water and other substances) into non-flowering and flowering groups. The record of the non-vascular plants is poor, whereas those with stronger tissues such as ferns, conifers and flowering plants are relatively common. Range Precambrian to Recent.

Algae

The single-celled blue-green algae may form resistant masses due to the agglutination of sedimentary particles on their sticky outer surfaces. These then form laminated bodies. Such organo-sedimentary structures are recorded from Precambrian and later sediments. They are termed stromatolites. The majority are found in shallow, tidal and subtidal environments and they may measure a metre (3 ft) or more in diameter. Smaller examples are common worldwide. Excellent examples of small algal stromatolites referred to the genus *Girvanella* may be collected from the Cambrian sediments of California. The circular laminated or layered appearance of the fossil is characteristic.

Other algae actually secrete calcareous support structures or 'skeletons'. These skeletal algae include the plate-like alga *Halimeda*. Usually the skeleton is broken down to form a calcareous sand, but sometimes it is preserved in the overall structure of the reef complex. Commonly known as the calcareous algae, genera such as *Halimeda* and *Goniolithon* are important rock formers. The remains of calcareous algae are known worldwide throughout the fossil record. *Halimeda* is a member of the green algae, the majority of which grow in shallow back-reef environments.

Some red algae also secrete calcareous skeletons. The majority of red algae occur within reef assemblages or along shorelines. The genus *Lithothamnion* is a typical representative of this group. It has an encrusting, globular or branched form that is made up of densely layered calcium carbonate. *Lithothamnion* and related genera are found in sediments from the Jurassic onwards.

Microscopic algae also secrete calcareous skeletons. The coccoliths are impossible to see with the naked eye but thick accumulations of their plates are evident worldwide. They include the famous white chalk of the White Cliffs of Dover, in southern England. The coccoliths are single-celled plants surrounded by protective calcareous platelets. They are planktonic and are known worldwide. The first coccoliths are recorded from sediments of Upper Triassic age.

Girvanella

5 cm

Stromatolite

Halimeda

Lithothamnion

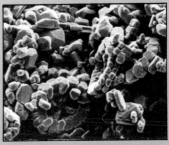

Coccoliths
(×2118)

Non-flowering plants

This description is here applied to the ferns, tree ferns, horsetails, conifers and related plants. The record of these plants dates from the Silurian to the present day.

Calamites This plant is a relative of the living horsetails. It has a jointed stem that is marked by strong vertical ridges. The jointed sections are shorter towards the growing tip. *Calamites* is particularly abundant in the deltaic sandstones of the Upper Carboniferous (Pennsylvanian).

Annularia *Annularia* and the related genus *Sphenophyllum* are also relatives of the living horsetails. Unlike *Calamites* they are best known through the preservation of leaf clusters. These develop as circlets around the jointed stem and they are often preserved as impressions on coal slabs. *Annularia* is known worldwide from Upper Carboniferous (Pennsylvanian) sediments.

Neuropteris *Neuropteris* and related genera, such as *Alethopteris* and *Pecopteris*, are fern-like plants. Some may even be true ferns. *Neuropteris* is known as a seed-fern, and it is difficult to separate its overall shape and form from that of a living fern. The leaves are compound and carry many small leaflets. A strong venation is present, with numerous veins arising from a distinct mid-rib. Unlike *Alethopteris* the veins are long and gently curved. *Neuropteris* is locally abundant in the Upper Carboniferous (Pennsylvanian) and is known throughout Europe, North America, North Africa and Asia.

Alethopteris *Alethopteris* is a coal measure seed-fern similar to *Neuropteris*. It has a multipinnate leaf and a characteristic venation. The leaflets are slimmer and straighter than those of *Neuropteris* and the veins shorter and less curved. *Alethopteris* is known throughout Europe, North America, North Africa and Asia.

Lepidodendron This is one of the best-known of all coal measure plants. Its remains may exceed 76 cm (30 in) in length and evidence of stems, leaves and roots are all recorded from coal measure sediments. Because the different parts were not always attributed to the same plant, the stem is named *Lepidodendron*, the roots *Stigmaria* and the leaves and branches *Sigillaria*. The stem is often massive and supports a crown of branches. The leaf scars that occur all over the stem are diamond-shaped or oval and are arranged spirally. *Lepidodendron* is known throughout Europe.

Glossopteris This is a seed-fern that gives its name to the distinctive flora that characterized Upper Carboniferous and Lower Permian times. It has a long narrow leaf with a delicate venation. *Glossopteris* flora is found throughout the southern continents and India.

Annularia
leaf

Neuropteris
leaf

HNV

Lepidodendron

Calamites stem

ANVO

Alethopteris

Stigmaria

5 cm

Glossopteris
leaf

Brachyphyllum   Conifers are an important group of living plants, and include the pines and redwoods. They first appeared during the Carboniferous and were particularly common during the Mesozoic and Tertiary periods. *Brachyphyllum* is a conifer found in Jurassic sediments. Remains consist of twigs with short leaves and male and female cones. It is cosmopolitan in its distribution.

Araucaria   *Araucaria* is a conifer native to the Southern Hemisphere. It is also cultivated as an ornamental tree in northern lands. *A. araucana* is commonly known as the 'monkey puzzle tree' whereas *A. heterophylla* is the 'Norfolk Island pine'. Species of *Araucaria* first appear during the Lower Mesozoic. They are known from leafy twigs, seed scales and compact globular cones.

Williamsonia   This plant is a representative of the extinct Bennettitales group of plants, the reproductive organs of which strongly resembled those of the flowering plants. *Williamsonia* had a robust stem and numerous frond-like leaves. The 'flowers' of *Williamsonia* have a discoid base comprising a number of petal-like bracts. They are solitary and held pollen on stamens that curved inwards and upwards. *Williamsonia* is known from the Upper Triassic of North America (e.g. Arizona and New Mexico) and the Jurassic of Europe (e.g. Yorkshire, England).

Sequoiadendron   This conifer includes among its various species the giant redwoods of California, USA. Some individuals of the species *Sequoiadendron gigantea* exceed 90 m (295 ft) in height. The fossil record of the redwoods began in the Lower Mesozoic and various species are known through their woods and cones. The cones are often small with a limited number of scales. *Metasequoia*, the dawn redwood, is a deciduous conifer. The species *M. occidentalis* is a well-known species from the Oligocene of the USA.

Ginkgo   *Ginkgo* is a representative of one of the oldest groups of non-flowering vascular plants. It is commonly known as the maidenhair tree and is a member of the order Ginkgoales. The first ginkgoes are recorded from sediments of Permian age. They have a broad, fan-shaped leaf which is invariably subdivided or lobed. The leaves have a parallel venation and occur in clusters at the end of short branches. The ginkgoes are locally abundant in Jurassic and Cretaceous sediments. They were once worldwide in their distribution but the living species, *G. biloba*, is endemic to North America and China. Ornamental trees are, however, found in many botanical gardens.

Brachyphyllum leaf

Araucaria cone leaf

Williamsonia leaf

Ginkgo leaf

Metasequoia leaf

5 cm

Flowering plants

The flowering plants, or **angiosperms**, constitute the most important group of plants. They first appeared during the Cretaceous and spread worldwide during the Cainozoic. They are divided on the structure of their seeds. In the **dicotyledons** the seed is comprised of two loosely joined cotyledons, whereas the seed of the **monocotyledons** consists of one component which stores food for the developing shoot. The leaves and wood of both groups differ significantly. In the dicotyledons the leaves have a network of veins and the vascular bundles are arranged in a single ring around the outer part of the stem. Monocotyledon leaves have parallel veins and the vascular bundles are scattered throughout the width of the stem.

Laurus This is the scientific name for the laurel. The leaf is long and its edge undivided or entire. A strong, central rib is present and the secondary veins diverge out from it. The laurel family were important during the Cretaceous. Distribution worldwide.

Acer Living varieties of this genus include the sycamore and the maple. The leaf is broad and three-lobed. The central rib is present in all three lobes and a network of smaller veins branch out over the surface of the leaf. *Acer* first appeared during the Palaeocene and is cosmopolitan in its distribution. It is locally abundant in Eocene and Oligocene sediments.

Nipa This palm is an example of a monocotyledonous angiosperm. It is characterized by large pear-shaped fruits. *Nipa* is found today in Malaysia, and its occurrence in the Eocene sediments of northern Europe is a clear indication that a tropical climate once prevailed. The vascular bundles are scattered throughout the stem of this palm. *Nipa* ranges from the Eocene to the present day and is known only from Europe and Asia.

Oncoba In contrast to the single-seed-bearing fruit of *Nipa*, the fruit of *Oncoba* is divided into several **locules**, each of which contains a seed. The fruit of *Oncoba* resembles a small melon with a warty skin. Internally the fruit consists of a mass of long stringy fibres. Recent relatives of *Oncoba* are found in tropical and subtropical environments. *O. variabilis* is a common fossil in the London Clay of Sheppey, England.

Fossil wood The remains of angiosperm stems, roots and branches are locally abundant in lake and inshore sediments during the Mesozoic and Tertiary. These remains are often silicified and the impregnated tissues preserved in fine detail. Fossil forests exist throughout Europe, North America and Africa with the fossil wood of trees such as the oak (*Quercus*) and the palm (*Palmoxylon*) among the most common. *Quercus* has growth rings whereas *Palmoxylon* does not.

Laurus leaf

Acer leaf

Nipa seed

Oncoba seed

5 cm

Quercus wood

Palmoxylon wood

Trace fossils

Trace fossils are the result of biological activity on or within a sediment. They take the form of tracks and trails, burrows and borings and even body waste deposits. Dinosaur footprints occur as distinct trails in ancient lake shore sediments whereas trilobite tracks criss-cross sedimentary rocks laid down on Palaeozoic sea floors. The shape or geometry of trace fossils varies with the needs of the animal that created them. Single vertical burrows indicate that the main requirement is for protection against predation or a rigorous environment, while complex horizontal patterns may be linked with a shortage of food and the need for an intensive 'farming' of the sea floor.

Skolithus   *Skolithus* is the name given to single, vertical tubes found in sands and sandstones deposited in shallow and very shallow water environments. The tubes may be 30 cm (12 in) in length and 2–4 cm (0.8–1.6 in) in diameter. They are associated with the burrowing habits of annelid worms. The Pipe Rock of Scotland is an excellent example of a sediment with *Skolithus*. Such traces are known worldwide since the Cambrian.

Monocraterion   This is a single-tubed burrow. It has a larger diameter than *Skolithus* and has a ringed structure that reflects the repeated working of the burrow by the animal. *Monocraterion* is typical of intertidal and subtidal environments. Excellent examples occur in the Jurassic calc-arenites of the Dorset coast of southern England.

Arenicolites   *Arenicolites* is a narrow U-shaped burrow. It is arranged vertical to the bedding plane and has small rounded to slightly fluted openings. Unlike *Monocraterion*, these burrows are not reworked internally and the ringed structures that reflect either movement of restructuring of the burrow are missing. *Arenicolites* is created by a shallow-water dweller. It is cosmopolitan in distribution and ranges from the Cambrian to the present day.

Ophiomorpha   This is an unusual burrow that may occur either singly or as a branched structure. The individual tubes are comparatively large, with a diameter of 3–6 cm (1.2–2.4 in). They have a knobbly surface texture and may terminate in a large bulbous structure. Such structures also occur at the junctions between branches. *Ophiomorpha* burrows are associated with crustacean activities. They are known throughout the Mesozoic and Cainozoic and recorded from most continents.

Borings   Bivalves such as *Pholas* and *Teredo* bore into rock or wood to create permanent shelters. They do so by a combination of mechanical and chemical processes; the shallow borings providing a protected niche. Borings are known throughout the Mesozoic and Cretaceous.

5 cm

Skolithus

Arenicolites

Monocraterion

Teredo

Ophiomorpha   *Pholas*

Thalassinoides   *Thalassinoides* burrows occur parallel or subparallel to the bedding plane. They are repeatedly branched with slight swellings at the junctions between offshoot tubes. The cylindrical burrows vary from 2 to 5 cm (0.8 to 2 in) in diameter. Some contain arthropod droppings, scratch marks and the remains of the crustacean that lived within. *Thalassinoides* are created by the ghost shrimp *Callianassa*. Such burrows are extremely abundant from the Jurassic onwards, worldwide.

Rhizocorallium   Like *Thalassinoides*, *Rhizocorallium* is a very common trace fossil. It occurs in shallow-water sediments and is particularly common in rocks of Jurassic age. *Rhizocorallium* is a U-shaped burrow in which the main tubes run parallel to each other. The tubes are between 2 and 4 cm (0.8 and 1.6 in) in diameter. Movement of the burrow through the sediment is reflected by the presence of lines or **spreiten** that mark the former position of the burrow bend. *Rhizocorallium* occurs in the horizontal or oblique plane. It is known in Mesozoic and Cainozoic strata and is cosmopolitan in distribution.

Coprolites and faecal pellets   These are fossilized droppings and body waste materials of animals. The term faecal pellet is used in association with the small droppings of gastropods or crustaceans, while coprolite is used to describe the fossilized faeces of crocodiles, dinosaurs and larger animals. Coprolites may exceed 30 cm (12 in) in length or width whereas faecal pellets rarely exceed 1 cm (0.4 in) in length. Fossilized droppings are known worldwide throughout the fossil record.

Cruziana   The name *Cruziana* is used in reference to the various tracks and trails formed by trilobites and trilobite-like arthropods. A typical track consists of two lobes which are the result of the animal scratching its way across the sediment. The traces are often long, and the lobes are parallel with a marked median depression. The scratch marks on the lobes were created by the walking part of the branched limb. Over 30 species of *Cruziana* have been identified. The elongate tracks are indicative of direct movement while searching of food or during migration. Heart-shaped traces, again bilobed and marked with scratch marks, represent temporary resting sites and were created by free-swimming animals. *Cruziana* tracks and trails are locally abundant in Lower Palaeozoic sediments, particularly of Cambrian and Ordovician age.

Pleurodictyon   This complex trace fossil is found in deep-water sediments, deposited in outer shelf and shelf slope environments. The trace consists of a number of small-diameter tubes arranged in geometric fashion. This takes the form of four or five-sided structures united in large numbers on the horizontal plane. Vertical, cylindrical tubes arise from the junctions between the horizontal tubing. It is thought that the main portion of the burrow system was formed below the sediment surface, with the vertical tubes used for access and irrigation.

5 cm

Thalassinoides

Coprolites

Rhizocorallium

Pleurodictyon

Cruziana

Precambrian Period

The discovery of fossils in sediments of Precambrian age is a rare occurrence. Algal stromatolites are the most common fossils found in Precambrian deposits. The oldest appeared 3.35 billion years ago. Fossil animals appeared much later, and the record of the first true fauna occurs in rocks between 680 and 600 million years old. This fauna was first discovered in the Pound Quartzite of Ediacara, in south Australia. It contains a number of soft-bodied jellyfish, hydrozoans and worms. Initially the fauna was thought to be confined in space and time, but further occurrences have now been recorded in Siberia, Europe, Canada and south-west Africa. The facing plate illustrates several genera from this famous community.

It is possible that the Ediacara Fauna is also our first true fossil **community**. For the biologist a community study will involve an analysis of the inter-relationships that exist between the various organisms that comprise the community itself. To the palaeontologist this type of study is restricted by the disappearance of most of the evidence. To the palaeontologist a community is, therefore, best defined as a group of organisms that lived in the same **habitat**. Naturally the palaeontologist can analyse the structure and function of a shell or skeleton and suggest modes of life for individual species. It is also possible to recreate the general structure of a community. An obvious example is a fossil reef, where the palaeontologist could rely on a modern-day equivalent for essential information. The study of the individual fossil in relation to its habitat is termed **palaeoautecology**, while the study of groups or assemblages of fossils is termed **palaeosynecology**.

A truly meaningful analysis of a community is restricted by the fact that the fossil record is incomplete, and that 70 per cent or more of an original marine community may have been soft bodied. Nevertheless, they represent a true, albeit incomplete, record of a fossil community.

The discovery of a highly fossiliferous bed may lead to the description of a fossil community. It is essential, therefore, to have some knowledge of the fossils that may occur in the same place at the same time. As stated above Precambrian communities are very rare, but from the Cambrian onwards the record of recurring assemblages greatly improves. Details of such assemblages and their ecology are clearly set out in the *Ecology of Fossils* edited by W. S. McKerrow.

The following pages describe and illustrate fossils found in rocks from the various periods of geological time. They do not necessarily occur in the same community but many are important representatives of a given assemblage. The discovery of one or two organisms may indicate that a number of others may exist under the same conditions, and should also be looked for. Where relevant, details of the assemblage are included in the description of the fossils of a given period. Once again it should be remembered that a full description of a fossil assemblage should include details of its relation to the sediment and a record of biological activity.

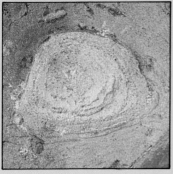

Stromatolites
(×2)

Medusina
(×2·1)

Cyclomedusa
(×2·1)

Dickinsonia

Spriggina

Cambrian Period

In the description of fossils from the Cambrian it should be remembered that the world was a very different place 580 million years ago. A great ocean named Iapetus separated Scotland and the north of Ireland from England, Wales and the rest of northern Europe, and North America was also divided. The rocks of Scotland have a greater affinity with those of Spitzbergen, Greenland and Eastern Canada than with the rest of Europe. In contrast sediments of eastern Newfoundland and Nova Scotia can be correlated with deposits of similar age in England and Wales.

Cambrian faunas are dominated by trilobites and inarticulate brachiopods, with the conical hyolithids (gastropod-like creatures) present in certain shelf sediments. The trilobites are useful stratigraphic tools with *Olenellus*, *Paradoxides* and *Olenus* representative of the Lower, Middle and Upper Cambrian divisions. The variety of trilobites found in Cambrian sediments is quite startling considering that no animals with mineralized skeletons are recorded from the Precambrian. Smallest of all are the trilobites known as agnostids and eodiscids. The latter are confined to Lower and Middle Cambrian rocks, while agnostid genera are recorded from the Upper Ordovician. Both stocks are characterized by equal-sized heads and tails and only two or three body segments. The eodiscids have small eyes and the agnostids were blind. Larger trilobites with small tails or spine-like telsons were common in the Lower Cambrian and, apart from *Olenellus*, *Redlichia* and *Paedeumias* are important genera. They are probably responsible for many of the traces found in shallow-water environments. In many areas the trace fossil *Cruziana* is found in association with the inarticulate brachiopod *Lingulella*. Together they represent a shallow water community that may also be characterized by ripple marks and other sedimentary structures. *Paradoxides* was also a shallow-water form and is often found in association with *Lingulella*, *Hyolithes* and the earliest orthid brachiopods in rocks of Middle Cambrian age. By the end of the Cambrian Period gastropods and graptolites had appeared and were important constituents of mid and outer shelf environments. In general terms the trilobites of the Upper Cambrian were less spinose. They had rounded headshields and tails and many belonged to families with either proparian or opisthoparian facial sutures. *Olenus* is a typical opisthoparian.

The graptolites are represented by **dendroid** genera such as *Dictyonema*. They were planktonic in habit and usually found in black shales and mudstones. Often they are pyritized. *Dictyonema* had a bell-shaped skeleton, the many branches supporting thousands of tiny animals in cup-like thecae. Bellerophontid gastropods scavenged on the sea floor. In life they had a well-defined head and foot, similar to those of the living slugs and snails.

Olenellus

Paradoxides

5 cm

Olenus

Lingulella

Dictyonema

Hyolithes
(×3)

Cruziana

Ordovician Period

The Iapetus Ocean persisted into the Ordovician, and the palaeogeography was similar to that of the Cambrian. Different groups of animals had appeared, however, and the trilobites were divided into distinct **faunal provinces**. These were centred on four areas, namely North America, northern Europe, southern Europe and Australia. *Ogygiocaris*, *Ampyx*, *Onnia* and *Bumastus* are among the many trilobites recorded from Ordovician sediments.

Of the new groups, various families of brachiopods arose and flourished during the Ordovician. Together with the orthids, strophomenids such as *Sowerbyella* and *Leptaena* were important constituents of the shallower water, **shelly** faunas. Bivalves had also appeared in numbers, although they would always be overshadowed by the brachiopods throughout the Palaeozoic. Gastropods, crinoids and bryozoans were well represented among sea floor communities.

The majority of trilobites were bottom crawlers. *Onnia* and related genera commonly referred to as **trinucleids** had large, semicircular headshields with long spines on the posterior edges. They had small tails and a limited number of body segments. The spines and large headshields supported the animal on the soft sediment surface. Many trinucleids were blind, others had small eyes. In contrast *Ogygiocaris* was much larger with the headshield and tail of roughly equal size. The eyes were crescent-shaped and the body rather smoother in overall shape. It is likely that *Ogygiocaris* was a more active crawler than the trinucleids. Trilobite tracks and trails are common in many Ordovician environments throughout the world. Trilobites are common constituents of most Ordovician assemblages.

Environments characterized by limestones or muddy limestones occur in the Lower Ordovician of North America and Scotland. Limestones are often associated with warm shelf seas and in the Ordovician the so-called carbonate communities are noted for the presence of algal stromatolites, brachiopods, nautiloids and trilobites. The indication is of very shallow conditions between low tide and 50 m (165 ft) deep.

In sandier sediments brachiopods and crinoids are more abundant. Strophomenids such as *Sowerbyella* and orthids including *Dalmanella* occur together with high-spired gastropods and crinoids such as *Balacrinus*. In these shallower sediments the shells are often fragmented on death by action of water and general transportation. Where this occurs the bedding plane or fossiliferous horizons will be characterized by shell debris. Whole and unabraded material will indicate limited transportation, and possibly the animals will be in their life positions.

Deeper water, black shales and mudstones from the Ordovician often contain abundant graptoloids. Sedimentation was probably slow and the action of bacteria limited. On death the planktonic graptoloids sank to the sea floor and were preserved intact. Throughout the Ordovician the graptoloids change in shape and size and are important for stratigraphic purposes. *Phyllograptus*, *Didymograptus* and *Orthograptus* are among the more common of Ordovician graptoloids.

Ogygiocaris

Ampyx

Onnia

Bumastus

Sowerbyella

Leptaena

Balacrinus

5 cm

Phyllograptus

Didymograptus

Orthograptus

Silurian Period

Although the Iapetus Ocean was narrower during the Silurian, distinct zoogeographical provinces still existed, as in the Ordovician and Cambrian. As before, the faunas of central Scotland and Northern Ireland were closer to those of Spitzbergen and Eastern Canada than to England, Wales and Nova Scotia. Different sedimentary environments have been recognized within these areas and diagnostic assemblages and communities described. Limestones are perhaps more common in England and Wales during the Silurian and the Wenlock Limestone of the Welsh borders is noted for its fossil riches. Bedding planes littered with brachiopod shells, bryozoans, crinoids and crinoid ossicles, tabulate corals, nautiloids, bivalves and trilobites indicate the former presence of a reef or reef slope. A decision on exactly which environment was involved would depend on evidence of the degree of fragmentation and transportation that had occurred. Of the brachiopods, *Leptaena* and *Atrypa*, a small spiriferid, are most common. The corals include *Favosites*, *Halysites* and *Heliolites*. Together they constituted, with the bryozoans, the main reef builders and binders. Crinoids would have flourished nearby. In life solitary rugose corals, sea-anemones and other organisms would have occupied surface niches and lived in crevices on the reef mass while active feeders such as fish, trilobites and nautiloids scavenged and fed on organic detritus. On death the free-swimming and loosely attached organisms would tend to be transported down the slopes. The coral heads in situ would indicate the position of the reef itself.

Brachiopods dominated the shallower water Silurian communities, spiriferids, strophomenids and rhynchonellids thriving in both limestone and sandy sediments. Shallow or fairly shallow-water brachiopods include *Lingula*, *Pentamerus* and *Sphaerirhynchia*. Trilobites such as *Dalmanites* are common in reef and outer shelf environments. In the latter they would be associated with orthid brachiopods and planktonic graptoloids.

The graptoloids of the Silurian Period are associated with deeper water sediments. They may be found in association with other invertebrates but mostly they occur in abundance on the closely packed bedding planes of black shales. Silurian graptoloids are mostly one-stiped forms. *Diplograptus* and *Rastrites* are examples of Silurian graptoloids.

At the end of the Silurian Period the Iapetus Ocean closed. Most of Europe, Greenland and much of eastern North America formed one huge land mass. Terrestrial deposits of the Upper Silurian contain evidence of the evolution of the first land plants and animals. These are extremely rare but their presence adds a new dimension to the collection of fossil materials.

5 cm

Favosites

Heliolites

Atrypa

Sphaerirhynchia

Pentamerus

Rastrites

Diplograptus

Devonian Period

Although large areas of the British Isles, northern Europe and north-eastern North America were dry land during the Devonian, lake, river and deltaic sediments contain evidence of new and exciting faunas and floras. Marine environments occur in south-west England, throughout Europe and North Africa and over much of the USA.

Freshwater deposits of Lower Devonian age (Old Red Sandstone) frequently contain fossil fish remains. Classic localities are recorded from Scotland, Wales, Gloucestershire in England, and from Canada. The fish are varied in shape and size but many of the Lower Devonian species belong to the jawless agnathans. *Cephalaspis* and *Pteraspis* are typical members of this group, with heavily armoured headshields and with a sucker-like mouth on the undersurface. They have thickly scaled bodies with relatively poorly developed fins and tail fins. These fish are frequently found in association with the predatory, jointed limbed eurypterids. Eurypterids usually lived in freshwater. They varied in size but some species measured more than 1 m (3 ft) in length. Swamp communities of the Devonian contained primitive vascular plants and early mites and spider-like arthropods. In later Devonian sediments bony fish, including lungfish, tended to replace the agnathans.

Reef communities and other shallow marine communities of Devonian age contain an abundance of invertebrate species. Stromatoporoids, corals and bryozoans are common in the limestones of Torquay, Devon, south-west England, and Ferques in northern France. The corals include the colonial tabulate *Thamnopora* and the colonial rugose genus *Hexagonaria*. These may occur as large masses. Associated fossils include orthocone nautiloids and the brachiopods *Meristina* and *Athyris*. In muddier limestones the trilobite *Phacops* is common. It had a rounded headshield and large compound eyes, many body segments and a small rounded tail.

Spiriferid brachiopods and goniatites are perhaps the most useful of Devonian fossils. They are found mostly in deeper shelf and basinal deposits. Both groups evolved rapidly during the Devonian. They are widely distributed and are used as **zone** or **index fossils**. The spiriferids include *Cyrtospirifer*. *Clymenia* is a representative goniatite.

Trace fossils are found in both freshwater and marine sediments of Devonian age. In freshwater and shallow marine sediments, usually sandstones, the burrows are either single or double vertical tubes. These may be attributed to the activities of small arthropods such as ghost shrimps and crabs. In deeper water marine sediments the activity may be more intense and related to extensive surface 'grazing'. Trilobites, gastropods and worms may be responsible for this churning of the sediments. The term **bioturbation** is used to describe sedimentary structures that result from such activity.

Cephalaspis
headshield

Thamnopora

Hexagonaria

5 cm

Meristina

Athyris
(×1·5)

Phacops

Cyrtospirifer

Clymenia

Lower Carboniferous (Mississippian) Period

The Carboniferous rocks of western Europe can be divided into two major subsystems. The lower subsystem is characterized by shelf limestones, the upper by sandstones, shales, mudstones and coal deposits. The limestones belong to the Dinantian Subdivision of the Carboniferous, and the clastic and carbonaceous rocks to the Silesian subdivision. Throughout Europe the Upper Carboniferous is termed the 'Age of Coal' as the vast coal deposits of several countries were laid down during this time. The two groups of rocks indicate very different environmental conditions. Shelf limestones are deposited in warm, clear seas rich in calcium carbonate. The sandstones and coal seams, however, originated in rivers, deltas and swamps. Where the limestones are overlaid by clastic sediments, this indicates the retreat of marine conditions.

In North America the Lower Carboniferous is termed the Mississippian. The rock types of this subsystem tend to be more varied than those of Europe, with sandstones and limestones deposited in different areas at approximately the same time. The Upper Carboniferous or Pennsylvanian is also noted for a variety of depositional environments. In Kansas, for example, coal deposits of Middle Pennsylvanian age are overlaid by Upper Pennsylvanian shelf limestone containing fusulinids and molluscs.

Carboniferous limestones in many parts of the world are noted for their reef and reef slope assemblages. Corals are abundant in many areas, with the colonial rugose genera *Lithostrotion* and *Lonsdaelia* and the tabulate *Syringopora* as examples. In some reefs algal stromatolites and bryozoans contributed much to the overall framework. Brachiopods such as *Productus* and the bivalve *Parallelodon* also occurred in numbers. Seawards from the reef, deposits rich in crinoids such as *Amphoracrinus* and blastoids such as *Petremites* occur. Spiriferids, productids, bivalves, small solitary corals including *Zaphrentis*, and the delicate lace bryozoans such as *Fenestella* are also common. In some outcrops the bedding planes are littered with the detritus of reef and reef slope dwellers. Accumulations of crinoid ossicles as crinoidal limestones are a clue to the transportation and disarticulation of skeletons. Fish including early sharks, nautiloids and goniatites represent the free-swimming animals of the Carboniferous seas. Small trilobites such as *Phillipsia* also occur in shallower water limestones. Lower Carboniferous limestones vary in colour and texture. Many are grey or grey-brown, while others are darker, blue-grey or dark grey. The latter are usually associated with either mud-rich environments or higher energy environments, depending on grain size. In some darker, coarser-grained limestones giant brachiopods such as *Gigantoproductus* may be found. This coarsely ridged form lived partially buried in the sediments, which is a niche occupied today by the giant clam.

Lithostrotion

Lonsdaelia

Syringopora

Amphoracrinus

Productus

Pentremites

Fenestella

Zaphrenitis

Phillipsia

Gigantoproductus

5 cm

Upper Carboniferous (Pennsylvanian) Period

Both marine and freshwater communities can be examined in sediments of Upper Carboniferous (Pennsylvanian) age. In areas such as Kansas algal limestones indicate shallow marine conditions while in Europe mudstones with goniatites and large thin-shelled bivalves were laid down in deep basinal waters. *Reticuloceras* is a representative goniatite and *Posidonia* a thin-shelled pteroid bivalve. Brachiopods may also occur in these mudstones.

During the latter part of the Upper Carboniferous marine bands, non-marine bands and coal deposits occur within repeated cycles of deposition. The cycle usually starts, as in central Kansas or in South Wales, with a non-marine sandstone. This is followed by a shale, a non-marine limestone, a clay band and a coal seam. Above the coal seam marine shales and possibly limestones occur. Bivalves such as *Carbonicola* are often common in the non-marine band, while goniatites such as *Gastrioceras* and *Reticuloceras* characterize the marine sediments. In Kansas, fusulinid foraminiferans are common in the marine units. The cyclic sedimentation of the Coal Measure Series is an indication of delta-swamp environments. In the swamps amphibians, fish and insects flourished as links in a complex food chain. The banks of streams and swamps were lined with ferns, lycopods and calamitids (relatives of the living horsetails). Fossil forests existed worldwide and the dead and decaying trunks of giant plants gave rise to coal we burn today. The ferns such as *Neuropteris* and *Alethopteris* gave a thick ground cover, while the calamitids and the gigantic lycopods formed the canopy. *Calamites* had a jointed stem similar to that of the living *Equisetum*. In contrast it was robust with prominent vertical markings. Individual plants grew 10 m (33 ft) high and their remains are common in certain sandstone horizons. The lycopod *Lepidodendron* reached 25 to 30 m (82 to 98 ft). It has a characteristic, diamond-shaped leaf scar and large specimens may be collected from both sandstones and coal seams. Fossil forests of Upper Carboniferous age have been excavated in Glasgow, Scotland, and Monceau les Mines in eastern France. At Monceau careful research and collection have resulted in the retrieval of many insects, amphibians and other fossils.

The Upper Carboniferous of Europe, and elsewhere in the world, marks the withdrawal of the sea from the land. The continents collided to form one land mass and arid terrestrial environments replaced the swamps and forests. In the Southern Hemisphere great ice sheets covered large areas of Australia, India, South America, southern Africa and Antarctica. Continental red-coloured beds and pebbly glacial mudstones are typical of the Permian world. The discovery and collection of fossils from Permo-Triassic deposits is a rare event, however, for reasons which are explained later.

Gastrioceras

Reticuloceras

Posidonia

Neuropteris

Carbonicola

Alethopteris

5 cm

Calamites

Lepidodendron

Permo-Triassic Period

For the purposes of this book the rather limited communities of the two periods that fall either side of the Palaeozoic-Mesozoic boundary are described under one heading. The limited nature of Permo-Triassic assemblages is related to palaeogeography (the distribution of land and sea). Throughout the Permian and for much of the Triassic the areas where we search for fossils were uplifted above sea-level. Terrestrial conditions prevailed with harsh arid, desert conditions existing over much of Europe, North and South America, and the southern continents. Red sandstones and mudstones are the predominant lithologies. The sandstones are often cross-stratified and dune-bedded.

During the Permian, restricted marine faunas and isolated reefs occur in north-east England and Germany. These are associated with the Zechstein Sea, which is best described as a warm, salt-rich sea that covered much of Germany, Poland and the North Sea Basin. In the Sunderland area of north-east England, reef communities are found in locally outcropping limestones. The reef debris contains the lace bryozoan *Fenestella*, the strophomenid brachiopod *Horridonia* and the bivalve *Parallelodon*. The bryozoans were the major frame builders, while the brachiopods and bivalves occupied surface niches. Back reef, lagoonal sediments contain nautiloids, gastropods and bivalves. In the salt-rich waters of coastal basins the bivalves were abundant. The Cannon Ball Limestone is a localized deposit of the Sunderland region.

In the western states of North America, marine conditions prevailed during the Permian. Reef build-ups are known from the Guadalupe Mountains of New Mexico. Fusulinid limestones are also characteristic of this region during Permian times. Abundant crinoids exist in the Permian limestones of the Pacific island of Timor.

The Triassic terrestrial conditions, coupled with widespread extinctions at the end of the Permian, result in a very poor fossil record for the period. Lagoonal sediments are recorded with a restricted invertebrate fauna of burrowing bivalves (*Pholadomya*) and brachiopods (*Lingula*). These occur in association with fish and reptile bones. Algal mat material also occurs in tidal deposits. Polished samples of the algal Cotham (Landscape) Marble of Somerset, England, make excellent collectors' items. The limestone is formed by blue-green algae and is a similar structure to the stromatolites of the Precambrian. Tracks, trails and burrows may be found in Triassic mudstones. In the Austrian Alps, near the town of Hallstadt, Triassic limestones have yielded beautiful examples of the ceratites *Joannites* and *Trachyceras*. At the end of the Triassic Period the continents began to split up and the sea covered many areas. Many new animals and plants evolved to occupy the niches left vacant at the end of the Permian.

Fenestella

Horridonia

Parallelodon

Pholadomya

Lingula

Joannites

Trachyceras

5 cm

Jurassic Period

In contrast to the Triassic, marine sediments of Jurassic age are widely distributed throughout Europe. They also occur in the western and central States of North America. It is possible in the British Isles to study outcrops of marine and coastal deposits and to collect many species of plants and animals previously unknown in the fossil record. The Lower Jurassic of England is characterized by the blue clays and thin muddy limestones of the Lias. These deposits were laid down in approximately 150–300 m (500–1000 ft) of water. The Lias communities varied in relation to the sediments on the sea floor, with the more silty areas supporting a more varied fauna.

Such sediments outcrop along the Yorkshire coast of England, and a rich fauna of ammonites, bivalves and crinoids may be observed on many bedding planes. Liassic ammonites include *Dactylioceras* and *Psiloceras*. *Pinna*, *Gryphea* and *Pecten* are among the more common bivalves. Trace fossils are abundant throughout the Jurassic. *Thalassinoides* and *Rhizocorallium* are the best known; both were formed by shrimp-like crustaceans. Brachiopods are abundant in Middle and Upper Jurassic sediments. Rhynchonellids such as *Goniorhynchia* are often associated with coarse sandy sediments. In northern France, near the town of St Aubin, coastal exposures of Jurassic sandstones and limestones yield thousands of rhychonellids and terebratulids. The fossils occur loose on the foreshore, trapped in crevices or partly buried in sand.

In the Middle Jurassic, clay deposits are common in many areas. Oysters and corals flourished in the lagoons in which these sediments were deposited. The corals include *Isastrea* and *Thamnasteria*. Oolitic limestones are also a common feature of the Middle Jurassic. These rocks were laid down in shallow warm seas, similar to those found today around the Caribbean Islands. They are often cross-stratified and the fauna is usually restricted to gastropods, bivalves and echinoids. These animals also occur in sandy limestones along with brachiopods and crustaceans. *Nucleolites* is a well-known echinoid from the Middle Jurassic.

Clays, sandstones, sandy limestones and reef limestones also characterize the Upper Jurassic. The major groups of fossils are well represented in most communities. The bivalve *Trigonia* is found in sandy limestones along with oysters and beautifully preserved trace fossils. At the end of the Jurassic the sea withdrew from areas of Europe and North America. Lagoon and freshwater swamp environments prevailed during this period and plants and vertebrates were the major constituents of the communities that existed within these environments. Dinosaurs, crocodiles and pterosaurs are known from both Europe and North America. *Archaeopteryx*, the first bird, is recorded from the Solnhofen Limestone of southern Germany. Upper Jurassic floras include horsetails, cycads and ferns. Their carbonaceous remains are common in the sandy sediments of the Yorkshire coast, England. Non-marine faunas and floras are also known from Colorado, Wyoming, Montana and other states of western North America.

Psiloceras

Dactylioceras

Pinna

Rhizocorallium
(×0·1)

Gryphaea

Pecten

Goniorhynchia

Isastrea

Thamnasteria

Nucleolites

Trigonia

Cretaceous Period

The development of coastal swamp and lake environments during the Upper Jurassic was continued in northern Europe in the earlier part of the Cretaceous Period. The 'Wealden' facies of southern England (Isle of Wight and Dorset) and Belgium contain plant and vertebrate remains. The latter includes the dinosaur *Iguanodon* which has also been found in North America.

Marine sands overlie the 'Wealden' facies and shallow-water communities, and contain corals, brachiopods, bivalves, gastropods and ammonites. Locally, as in the Lower Greensand of the Isle of Wight, England, a variety of crabs, lobsters and prawns occur. They include the genus *Hoploparia* which may have been responsible for some of the extensive burrowing found within the sediment. Certain bivalves are often abundant in these sediments.

Clays and silty clays were also deposited during the Lower Cretaceous. The Gault Clay of northern Europe is noted for its grey-green colour and for a rich fauna of ammonites and bivalves. Of the ammonites *Hoplites* and *Anahoplites* are among the most common. Their preservation is often exceptional and enhanced by pyritization. Burrowing bivalves such as *Nucula* and *Trigonia* (*Linotrigonia*) occur, together with surface dwellers such as *Spondylus* and *Inoceramus*. Belemnite guards, common in many clays of Jurassic and Cretaceous age, are represented by the genus *Neohibolites*.

With the passage of time the sediments of the Lower Cretaceous fluctuate between sands and clays. These represent different environmental conditions that support different groups of organisms. In the Upper Cretaceous the trend is towards deep-water conditions, and fine-grained limestones spread over vast areas of Europe and North America. In Europe the chalk marls and chalk of south-east England and northern France are such deposits, while in North America the Niobrara chalk of Kansas represents similar depositional conditions. Ammonites, bivalves and brachiopods are probably the most abundant elements of these chalky sediments. The brachiopods are represented by terebratulids and rhynchonellids. These lived on the surface of the sea floor whereas bivalves such as *Lithophaga* and *Cucullaea* bored or burrowed into the substrate. The ammonite fauna of the chalk seas was very varied with coiled, uncoiled and spiral forms occupying different niches. *Schloenbachia*, *Homites* and *Turrilites* are also abundant at certain levels. The sponges include *Ventriculites*, a rooted, vase-shaped hexactinellid. Echinoids such as *Micraster* and the free-swimming crinoid *Marsupites* provide important zone fossils for the Upper Cretaceous. In North America the chalks and chalky marls of Kansas are rich in vertebrate remains, with turtles and mosasaurs among the most common finds. In southern states, such as Texas, rudist reef limestones outcrop. These can be correlated with similar deposits in Spain, France and north Africa. They mark the shoreline of the ancient equatorial ocean named Tethys.

Hoploparia

Hoplites

Spondylus

Nucula

Inoceramus

Neohibolites

Lithophaga

Micraster

Ventriculites

Turrilites

Marsupites

5 cm

Cainozoic Period

The continued separation of the continents throughout the Mesozoic resulted in many geographic changes, and generated the mountain-building episodes that resulted in the Alps and Himalayas. The chalk seas retreated, and sands and clays replaced the fine-grained limestones of the Upper Cretaceous throughout northern Europe and along the eastern coast of North America. In the Tethyan realm of southern Europe and the southern states of the USA limestones persisted, but the rudists characteristic of the Cretaceous had vanished. This is true of many Mesozoic species, both plant and animal, as the end of the cretaceous is marked by an episode of worldwide extinctions. Ammonites, dinosaurs and many other groups had by this time died out and the Cainozoic is often termed the Age of Mammals. It is also the era during which bivalves and gastropods dominate sea-floor communities and the flowering plants terrestrial floras.

In the Mediterranean region, formerly part of the Tethys Sea, limestones formed of nummulites outcrop as resistant scarps. The nummulites thrived in warm, shallow seas, and their presence in northern Europe is restricted to the Paris Basin, France, and the Hampshire Basin, England. In these areas sands and clays are the most common types of rock. They represent both marine and freshwater environments, and indicate that the Cainozoic shoreline fluctuated back and forth across the edges of the continents. In the marine sands gastropods such as *Aporrhais*, *Athleta* and *Turritella* are often abundant. This is also true of many species of bivalve, with banks of individual species outcropping in many localities. In south-east England the large clam *Venericardia planicosta* is particularly abundant in the glauconitic sands and sandy clays that outcrop on the coast of Sussex. *Venericardia* is also known from the Eocene deposits of Texas, USA. Fish, particularly sharks, are also common in Cainozoic sands and clays, with various species of *Odontaspis* recorded worldwide. Individual species are identified solely on their sharp pointed teeth; the sharks have a cartilaginous skeleton and this decays quickly on death.

In the marine clays, bivalves and gastropods are again the dominant faunal elements. The bivalves vary with water depth with *Nucula* and *Nuculana* present in deeper water sediments. *Pinna* and *Artica* are found in shallow marine clays of Eocene age. Oysters are also abundant in shallower water clays and in more sandy estuarine deposits. Fossil wood, infested by the shipworm *Teredo*, is frequently found in marine clays.

Both sands and clays may be extensively burrowed. *Thalassinoides* is found in many horizons, its presence in clays being picked out by the presence of septarian nodules. These are fine-gained, calcareous nodules which exhibit desiccation fractures on the inside. They form within the clays, and the remains of *Thalassinoides* and other traces occur as casts over their outer surfaces. The nodules are often fractured internally, the fractures lined with calcite crystals. Some septarian nodules form around the remains of crabs or lobsters while others contain fish, turtles or crocodiles.

Aporrhais

Athleta

Nummulites

Turritella

Venericardia

Odontaspis
teeth

Nuculana

Teredo

Pinna

Thalassinoides
(×0·06)

5 cm

Bibliography

Brazier, M. D. *Microfossils*. London: George Allen and Unwin, 1980.

Clarkson, E. N. K. *Invertebrate Palaeontology and Evolution*. London: Allen and Unwin, 1979.

Hamilton, W. R., Woolley, A. R., and Bishop, A. C. *The Country Life Guide to Minerals, Rocks and Fossils*. Feltham, England: Newnes, 1985.

McKerrow, W. S. (Editor). *The Ecology of Fossils*. London: Duckworth, 1978.

Moody, R. *Hamlyn Nature Guides: Fossils*. Twickenham, England: Hamlyn, 1979.

Moore, R. C. (Editor). *Treatise on Invertebrate Palaeontology*. Lawrence, Kansas: University of Kansas Press.

Murray, J. W. (Editor). *Atlas of Invertebrate Macrofossils*. London: Longman, 1985.

Romer, A. S. *Vertebrate Palaeontology*. Chicago: University of Chicago Press, 1974.

Tasch, P. *Palaeobiology of the Invertebrates*. New York: Wiley, 1973.

Whitten, D. G. A. and Brooks, J. R. V. *The Dictionary of Geology*. Harmondsworth, England: Allen Lane, 1978.

Glossary

Aboral Describes the upper surface of a sea-urchin test, on which are found a cluster of small plates – the periproct around the anus.

Adoral Describes the lower surface of a sea-urchin test, on which is found the mouth.

Anterior Front or forward portion of an animal or its skeleton.

Apical angle Angle formed across the outer edges of the initial whorls and tip of a gastropod shell.

Assemblage zone Stratigraphic division characterized by a fossil assemblage (group of fossils).

Astrorhiza(e) Small tube or channel-like structure that occurs in star-shaped clusters on the surface of a stromatoporoid skeleton.

Bedding plane Planar feature or surface within a sedimentary rock that parallels the original depositional surface on the sea floor.

Bifurcate Describes the branched nature of a rib on the surface of a cephalopod shell.

Biserial Term used to describe the distribution of thecae on both sides of a graptolite stipe or branch.

Brachial valve One of two valves which make up the shell of a brachiopod; bears support structure for internal feeding organ.

Bract Small leaf with a flower growing at its axil; the point at which the leaf rises from the stem.

Byssus Thread-like attachment found in bivalves; used to attach shell to sea floor or to other organisms.

Camera(e) One of many chambers found within the shell of a cephalopod. Chambers are separated by partitions (septa) and filled with gas.

Carapace Upper shell unit of tortoises and turtles; also the outgrowth that covers the head region of various crustaceans.

Cheilostome Type of bryozoan in which calcareous chambers are short and swollen or sac like, with a hinged lid.

Clast Fragment of rock (lithoclast) or fossil (bioclast) material found in sedimentary rock.

Concavo-convex Describes a brachiopod shell in which brachial valve is concave, and pedicle valve convex.

Corallum Mineralized skeleton of a coral colony.

Costa(e) Rib-like, thickened ridge on brachiopod shell that extends forwards from beak area.

Crenulate Having the edge divided into a small, tightly folded shape.

Cyclostome Variety of bryozoan consisting of simple, undivided calcareous tubes with rounded apertures.

Denticulate Having a tooth-like appearance.

Dissepiment Small, often curved vertical plate that develops inside the boundary wall of the coral skeleton.

Epitheca (coral) Outer wall of a coral skeleton.

Evolute Describes a shell in which successive coils are in contact, but do not overlap.

Facies Distinctive set of characteristics that occur within a given rock, eg grain size or texture.

Flexure Fold or bend in rocks.

Foramen Bounded opening found at or near the beak of the pedicle valve of many brachiopods.

Gonatoparian Describes those trilobites in which the facial suture bisects the genal angle.

Gyrocone Curved, slender type of shell found in some nautiloids.

Heteromorphic Term used to describe a variety of uncoiled ammonoids of Cretaceous age.

Horizon Bed of sedimentary rock with characteristic fossils.

Involute Term used to describe a shell in which coils are overlapped by later ones.

Inter-radial One of many small plates that occur between the 'rays' of a crinoid cup.

Irregular echinoid Variety of sea-urchin in which the mouth and anus may have migrated away from the central position.

Keel Ridge-like feature found on outer whorl of some ammonoid shells.

Lamella(e) Thin sheet of calcite or aragonite, characteristic of various bivalve shells.

Lithology Term used to describe the overall nature of a rock, based on its texture, grain size and composition.

Ligament pit Depressed area along the hinge line of bivalves that in life houses the ligament which opens the valves.

Locule Small chamber that contains a seed in higher plants.

Mamelon A mound-shaped structure found on the surface of stromatoporoid colonies.

Neural arch Arched structure of bone that occurs above the central mass of a vertebra. It surrounds the hollow neural canal through which passes the spinal cord.

Oncolite Banded clast (fragment) of algal origin.

Operculum (coral) A lid-like structure found in the 'slipper-coral' *Calceola sandalina*.

Opisthoparian Describes those trilobites in which the facial suture cuts the rear margin of the headshield in from the genal spine or angle.

Orbit Opening for the eye within the skull.

Orthocone Straight, slender type of shell found in certain nautiloids.

Oscule (osculum) Large opening on upper surface of sponge for outlet of water.

Pedicle valve One of two valves which make up the shell of a brachiopod; houses the pedicle.

Planispiral Simplest type of coiling found in gastropods; midline of shell occurs in single plane.

Plano-convex Describes brachiopod shell in which brachial valve is flat, and pedicle valve convex.

Phragmacone Chambered portion of a belemnoid skeleton.

Polyp Soft, flexible body of a sea-anemone or coral; flattened on lower surface with mouth and tentacles at top.

Posterior Back or rear portion of animal or its skeleton.

Proparian Describes those trilobites in which the facial suture cuts the lateral margin of the headshield.

Pro-ostracum Lip-like projection of the belemnoid phragmacone, the chambered portion of the skeleton.

Radial One of five identical plates that form the upper circlet of plates in the crinoid calyx.

Regular echinoid Echinoid in which the mouth and anus occur in the centre of the adoral and aboral surfaces.

Rhabdosome Name given to complete graptolite colony.

Rostrum A dense, bullet-like structure, also called the guard, which forms the major part of the belemnoid skeleton.

Scute One of a number of horny plates that cover the bony mozaic of a turtle shell.

Septum Vertical calcareous wall or plate found in corals, and in the mid-line of certain brachiopods.

Sinus Indentation found along the pallial line of many bivalve shells.

Siphon Fleshy tubes found in bivalves and gastropods, used for intake and outflow of water.

Siphuncle Tube in cephalopod shell that extends from living chamber to initial coil, piercing the mid-region of the septa.

Spiracle Small opening found in each of the deltoid plates of the blastoid theca.

Spondylium Curved, plate-like structure for muscle attachment found in the pedicle valve of articulate brachiopods.

Striae Scratches or tiny grooves on the surface of a rock.

Strophic Term applied to brachiopods in which the width across the hinge line is equal to or greater than the maximum width of the shell.

Sulcus Major depression found usually on the pedicle valve of some brachiopods.

Suture Line that marks the contact between the outer wall and internal septum in cephalopod shells.

Tabulae Horizontal element that divides the skeleton of various corals; may be curved or flat.

Tegmen Protective outer covering found in certain plants or animals.

Telson Spine found towards the rear of the axis in certain trilobites; considered to be equivalent of the pygidium.

Theca Small cup or tube found along the edge of graptolite stipes or branches.

Tubercle Small, rounded projection or outgrowth on the body surface of various animals or plants.

Type genus Typical and representative genus of a given family.

Uniserial Term that describes the occurrence of thecae along only one side of a graptolite stipe or branch.

Vascular bundle Bundle of vessels (xylem and phloem) that pass from root end to tip of stems and leaves, carrying water and dissolved food material.

Venter Outer area of shell or shell whorl in cephalopods.

Ventral Lower or under surface.

Invertebrate structures

The following illustrations will help the reader to understand something of the structure of many of the invertebrates which may be encountered as fossils, as well as helping to explain some of the terms used in the text which accompanies the identification plates. Other terms are described in the glossary.

Remember that you will sometimes find the original shell or body of the specimen, but in the case of specimens from very ancient environments the fossil may be replaced by minerals such as calcite or pyrite. All these factors will cause your specimen to take on an altered appearance. Look at the fossil from various angles – it may be just the tip of a shell that you have, or a section through a coral, for instance. Sometimes only an impression of the fossil will be left, in the form of a mould or cast. Also, in many cases pores and other structures will have been filled in, again making the appearance of the fossil less obvious than sometimes depicted.

Sponge   **Coral**

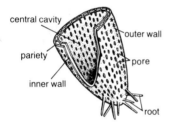 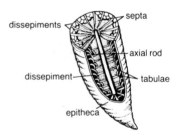

Mollusc
(gastropod; shell partly cut away)

Mollusc (ammonite *below*,
belemnite guard *bottom*)

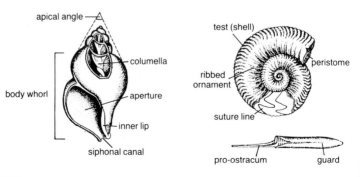

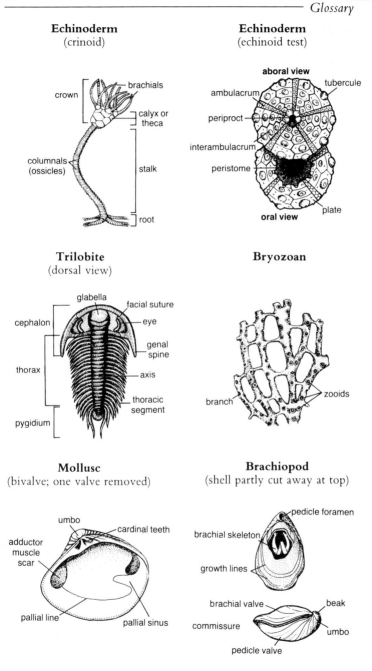

Echinoderm
(crinoid)

crown — brachials
calyx or theca
columnals (ossicles)
stalk
root

Echinoderm
(echinoid test)

aboral view
ambulacrum — tubercule
periproct
interambulacrum
peristome
oral view — plate

Trilobite
(dorsal view)

glabella
facial suture
cephalon
eye
genal spine
thorax
axis
thoracic segment
pygidium

Bryozoan

branch — zooids

Mollusc
(bivalve; one valve removed)

umbo
cardinal teeth
adductor muscle scar
pallial line
pallial sinus

Brachiopod
(shell partly cut away at top)

pedicle foramen
brachial skeleton
growth lines
brachial valve — beak
commissure
umbo
pedicle valve

Index

Page numbers in italic refer to illustrations